FOOD MICROBIOLOGY

ABOUT THE AUTHOR

Dr. Dilpreet Tulsi, M.Sc. Ph.D. in Microbiology is Asst. Professor on Adhoc basis at J.D. Verma Mahavidhyala, Manikpur under J.P. University. Dilpreet is teaching in this college for last four years. He is also a visiting faculty at reputed IAS Academy. He has attended three national and International level seminars. He has also completed a minor research project under UGC. Dilpreet has published his two articles also. Dilpreet is a man with a passion for rural development especially in the area of eradication of rural poverty, using dairy as an accelerator.

ABOUT THE BOOK

Food microbiology is the study of the microorganisms that inhabit, create, or contaminate food. "Good" bacteria, however, such as probiotics, are becoming increasingly important in food science. In addition, microorganisms are essential for the production of foods such as cheese, yogurt, other fermented foods, bread, beer, and wine.

Food safety is a major focus of food microbiology. Pathogenic bacteria, viruses, and toxins produced by microorganisms are all possible contaminants of food. However, microorganisms and their products can also be used to combat these pathogenic microbes. Probiotic bacteria, including those that produce bacteriocins, can kill and inhibit pathogens. Alternatively, purified bacteriocins such as nisin can be added directly to food products. Finally, bacteriophages, viruses that only infect bacteria, can be used to kill bacterial pathogens. Thorough preparation of food, including proper cooking, eliminates most bacteria and viruses. However, toxins produced bycontaminants may not be heat-labile, and some are not eliminated by cooking.

FOOD MICROBIOLOGY

Dr. Dilpreet Tulsi

WESTBURY

WESTBURY PUBLISHING LTD.

ENGLAND (UNITED KINGDOM)

Food Microbiology
Edited by Dr. Dilpreet Tulsi
ISBN: 978-1-913229-95-5 (Hardback)

Published by **Westbury Publishing Ltd.**
Address: 6-7, St. John Street, Mansfield,
Nottinghamshire, England, NG18 1QH
United Kingdom
Email: - info@westburypublishing.com
Website: - www.westburypublishing.com

British Library Cataloguing in Publication Data:
A catalogue record for this book is available from the British Library.

For more information regarding Westbury Publishing Ltd and its products, please visit the publisher's website- **www.westburypublishing.com**

Preface

Food microbiology is the study of the microorganisms that inhabit, create, or contaminate food. "Good" bacteria, however, such as probiotics, are becoming increasingly important in food science. In addition, microorganisms are essential for the production of foods such as cheese, yogurt, other fermented foods, bread, beer, and wine.

Food safety is a major focus of food microbiology. Pathogenic bacteria, viruses, and toxins produced by microorganisms are all possible contaminants of food. However, microorganisms and their products can also be used to combat these pathogenic microbes. Probiotic bacteria, including those that produce bacteriocins, can kill and inhibit pathogens. Alternatively, purified bacteriocins such as nisin can be added directly to food products. Finally, bacteriophages, viruses that only infect bacteria, can be used to kill bacterial pathogens. Thorough preparation of food, including proper cooking, eliminates most bacteria and viruses. However, toxins produced bycontaminants may not be heat-labile, and some are not eliminated by cooking.

Fermentation is one way microorganisms can change a food. Yeast, especially Saccharomyces cerevisiae, is used to leaven bread, brew beer, and make wine. Certain bacteria, including lactic acid bacteria, are used to make yogurt, cheese, hot sauce, pickles, fermented sausages, and dishes such as kimchi. A common effect of these fermentations is that the food product is less hospitable to other microorganisms, including pathogens and spoilagecausing microorganisms, thus extending the food's shelf-life. Some cheese varieties also require molds to ripen and develop their characteristic flavors.

– Author

Contents

1

Food Preservation with Chemicals

Some Considerations

- What organisms are the target?
  - Bacteria (procaryotes) vs. yeasts or molds (eucaryotes)
  - Gram-negatives vs. Gram-positives
- What are the properties of the food system?
  - pH
  - Lipid, protein content
  - Natural inhibitors
- How effective is the preservative in food vs. model systems?
- What concentration is needed?
- Is the preservative safe (GRAS)?
- Is it only allowed in certain foods or at certain levels?
- Is it natural or synthetic?

Mode of Action

- In general terms:
  - Killing vs. growth inhibition
  - Specificity for certain organisms or groups
  - Inhibition of spore germination or outgrowth
- More specific effects on cells:
  - Inhibition of nutrient transport
  - Enzyme inhibition

- Acidification of cell cytoplasm
- Membrane effects
  - Disruption of proton motive force
    a. Undissociated acids transport protons across membrane
    b. Proton released in cytoplasm
    c. Cytoplasm is acidified
    d. Alteration of membrane permeability
    e. Uptake inhibited, or leakage of intracellular compounds

Lipophilic Compounds

- Preservatives in this group
  - Benzoic acid and benzoates
  - Parabens
  - Sorbic acid and sorbates
  - Propionic acid and propionates
- Characteristics
  - Can pass through cell membrane
  - More effective at low pH
    a. In undissociated (uncharged) form
    b. Additive effect of low pH
  - Used to control yeasts and molds (bacteria already inhibited)
  - Used in foods susceptible to yeast and mold spoilage
  - Allowable limits ~0.1-0.2%
- Action
  - Inhibit substrate uptake
  - Carry protons into cell
  - Increase energy output of cells

Hydrophilic Compounds

- Preservatives in this group
  - Sulfur dioxide and sulfites
  - Nitrites and nitrates
  - Sodium chloride
  - Sugar
  - Acids
- Characteristics

  - Small molecules
  - Interact with cell components
  - Affected by pH, oxidation/reduction potential
  - May affect available oxygen
  - Affect bacteria as well as yeasts and molds
- Action
  - Aulfites: mechanism unknown
    a. pH of food is lowered
    b. Available oxygen reduced
    c. Essential enzymes inhibited
  - Nitrite: both oxidizing and reducing agent
    a. Key to color development in cured meats
    b. Interferes with key iron-sulfur enzymes
    c. Blocks ATP production from pyruvate
    d. Inhibits vegetative cell growth and spore germination
  - Salt and sugar
    a. Tie up available water
    b. Water lost from cells to equilibrate solute concentration
    c. Not affected by pH
    d. Salt more inhibitory than sugar
    e. Acids (lactic, acetic, citric, malic, tartaric, succinic)
    f. Lower pH of environment
    g. Acidify cytoplasm if they can enter cell
    h. May chelate ions

Bacteriocins and Antibiotics

- How are these different?
  - Bacteriocins are:
    a. Peptides, proteins, or protein complexes
    b. Synthesized by bacteria
    c. Inhibitory to (usually) closely related bacteria
    d. Membrane-active (disrupt permeability, PMF)
  - Antibiotics are:
    a. Secondary metabolites, not necessarily proteins
    b. Synthesized by many organisms, especially molds and streptomycetes

c. Inhibitory to a wide range of organisms

d. Active at various targets in the cell

- Concerns about antibiotics
- Selection of antibiotic-resistant microorganisms in foods
- Transfer of resistance genes among organisms
- Development of resistance to antibiotics used therapeutically
- Effect of antibiotic residues on humans

- Nisin
  - Only bacteriocin approved for food use in U.S. (processed cheese)
  - Possible adjunct for heat processing
  - Inhibits spore germination
  - Alters membrane permeability
- Bacteriocins as natural food preservatives
  - Produced by GRAS organisms (lactic acid bacteria)
  - Entire culture can be dried down and added to food
  - Mostly affect Gram-positive organisms, at present
  - Rapidly bactericidal
  - Other antimicrobials produced:
    a. Lactic acid
    b. Hydrogen peroxide

Other Chemicals

- Indirect antimicrobials
  - Added for other purposes but have antimicrobial effect
  - A flavoring agents more antifungal than antibacterial
  - Fatty acids and esters probably combine with other chemicals
- Ethylene and propylene oxides
  - Exist as gases and used as fumigants
  - Alkylate reactive groups in proteins, also DNA
  - Sterilizing agent for packaging
- Naturally occurring enzymes and proteins
  - Conalbumin (egg white) binds iron
  - Avidin (egg white) binds biotin
  - Lysozyme (egg white) degrades Gram + cell walls
  - Lactoferrin (milk) binds iron
  - Lactoperoxidase (milk) combines with H2O2; damages membranes

- Miscellaneous chemicals
  - H2O2 sterilizes surfaces, oxidizes
  - Ethanol dries and denatures
  - Polyphosphates chelate divalent ions
  - Wood smoke contains formaldehyde and other inhibitory organic compounds
  - Diethylpyrocarbonate kills yeasts; banned in U.S. because of possible carcinogen formation
- Antimicrobial Chemicals
  - Lipophilic molecules
    a. Sorbic acid CH_3-$(CH)_4$-COOH
    b. Propionic acid CH_3CH_2COOH
  - Hydrophilic molecules
    a. Sulfur dioxide SO_2 sodium nitrite $NaNO_2$
    b. Sulfite SO_3= sodium nitrate $NaNO_3$
    c. Bisulfite HSO_3- salt NaC_l
    d. Metabisulfite S_2O_5= sugars $C_6H_{12}O_6$
    e. Acids RCOOH
  - Bacteriocins and antibiotics
    a. Nisin natamycin
    b. Subtilin tetracycline
  - Other chemicals
    a. Antioxidants: BHA, BHT, TBHQ, PG
    b. Flavoring agents: essential oils, spices
    c. Emulsifiers and surface-active agents: fatty acids and esters
    d. Enzymes and proteins: lactoperoxidase, avidin, lysozyme, lactoferrin
    e. Sterilizing agents: ethylene and propylene oxide, hydrogen peroxide

PRESERVATION BY RADIATION

Properties of Radiation

- Terms
  - Ionizing radiation (<2000 A) : ionize molecules in its path
  - Curie: quantity of radioactive substance (now becquerel)
  - Roentgen: unit of measure for exposure dose of radiation

- Rad: unit of measure for absorption of radiation energy 1 krad = 1000 rads 1 gray = 100 rads
- ev: energy gained by one electron moving through 1 volt

• Radiation of interest
  - UV
    a. Nonionizing, but bactericidal
    b. Absorbed by proteins and à nucleic acids à mutations
    c. Poor penetration à limited to surfaces
  - Beta rays
    a. Stream of electrons
    b. Produced by linear accelerator
    c. Poor penetration (but better than UV)
  - Gamma rays
    a. Emitted from excited nucleus (60Co, 137Cs)
    b. Excellent penetration
  - X rays
    a. Similar to Gamma rays
    b. High-speed electrons bombard heavy metal
  - Microwaves
    a. Long à wavelengths ànonionizing
    b. Rapidly alternating field
    c. Heat generated by molecular friction of oscillations

Radiation vs. Microorganisms

• Factors controlling effectiveness
  - Types of organisms
    a. Gram-positive more resistant than Gram-negative
    b. Sporeformers more resistant than nonspore-formers
    c. Yeasts and molds more resistant than bacteria
    d. Extremely resistant organisms exist
  - Number of organisms: larger number à less effective
  - Age of organisms: mostresistant in lag phase
  - Composition of medium: proteins are protective
  - Oxygen level: higher resistance in anaerobic environment
  - Physical state:higher resistance in dry or frozen state

• Why are there such differences in resistance?

- Consider effects of radiation:
  a. Rapidly alternating field
  b. Generate ionized and electrically excited molecules
  c. Radiolysis (splitting) of water molecules
  d. Free radical and peroxide formation
  e. Damage to DNA
- Organisms can vary in:
  a. Cell wall and cell membrane composition
  b. Enzyme systems to quench free radicals, peroxides
  c. Content of reducing compounds
  d. DNA repair systems
- Rsafety in numbers phenomenon
- Nongrowing cells are not replicating DNA
- Availability of compounds to quench free radicals
- Availability of oxygen for peroxide formation
- Less water available in dry or frozen state

• Extremely radiation-resistant organisms:

Deinococcus	Rubrobacter
Deinobacter	Acinetobacter
viruses (small targets)	spores (esp. C. botulinum)

Radiation of Foods

• Sources:
- Gamma rays from Cs or Co source
- Electron beam from linear accelerator

• Treatment levels
- Radappertization: * commercial sterilization - 30-40 kGy * harshest treatmentàchem. changes
- Radicidation: * pasteurization - 2.5-10 kGy * kill Salmonella in raw poultry * treat spices
- Radurization: * "mild" pasteurization - 0.75-2.5 kGy * reduce spoilage organisms * extend shelf life of produce * kill insect eggs, larvae, etc.

• Effects in Foods
- Chemical changes with higher levels of irradiation:
  a. Off-flavors from free-radical, peroxide reactions

b. Lipid oxidat ion
c. Release of NH_3, H_2S from proteins and nitrogenous cpds
d. Partial destruction of B-vitamins
e. Pectin and cellulose degradation→softening
f. Enzymatic changes

– Reduce effects by:

a. Reduce temperature→slow free radical movement
b. Reduce oxygen→fewer oxidative free radicals
c. Add free radical scavengers
d. Remove off-flavors

- Foods approved for irradiation by U.S. FDA:

Food	Purpose	Dose Limit
fruits and	slow growth	up to 1 kGy
vegetables	and ripening	control insects
dry herbs, spices	kill insects	up to 30 kGy
seeds, teas,	control	0.3 kGy - 1 kGy
seasonings	microorganisms	
pork	control Trichin-	
	ella spiralis	
White potatoes	inhibit sprout	50 - 150 Gray
	development	
Wheat, wheat	control insects	200 - 500 Gray
flour		

Low Temperature Processes

Basic Principle

- Slow down or stop activities of microorganisms
- Not a lethal process (but some viability loss may occur)
- Q10 = ratio of reaction velocities at temperatures 10oC apart
  – For most enzyme-catalyzed reactions, Q10 = 1.5-2.5
  – Reaction rates decrease by ~2x with each 10o fall in temp.

Temperature Ranges

- Chilling = 10-15°C - fruit and vegetable storage
- Refrigeration = 0-7°C - for perishable and semiperishable foods
- Freezing = below -18°C - can still get microbial growth

Physical Effects at Low Temperature

- Relative humidity generally higher at chill temperatures

- Reduced aw as available water freezes
- Oxygen solubility increases with decreasing temperature
- Some pH changes may occur

Important Organisms

- Psychrophiles and psychrotrophs can grow at low temp.
- Low-temperature spoilage organisms mostly Gram-negatives:
  - Pseudomon as Acinetobacter
  - Alcaligenes Flavobacterium
- Many yeasts and molds can grow at low temp
- Most pathogens don't grow below 6C, but exceptions occur:
  - Listeria monocytogenes
  - Vibrio parahaemolyticus
  - Aeromonas hydrophila

Freezing

- First blanch in hot water or steam to:
  - Inactivate enzymes
  - Reduce microorganisms
  - Fix color
  - Wilt leafy vegetables
  - Displace entrapped air
- Quick (-20°C in 30 min) vs. slow (-20°C in 3 to 72 hr) freezing: small vs. large ice crystals time of exposure to concentrated solutes opportunity to adapt % halted vs. continued metabolism

Effects on Microorganisms

- Population level some cells killed during freezing process number of survivors will decline upon storage viability loss is greatest close to freezing point don't expect complete lethality
- Cellular level
  - Free water freezes; bound water doesn't
  - Cells dehydrated as free water reduced
  - Solutes concentrated in remaining available water
  - Viscosity of cytoplasm increased
  - Gases (O_2 and CO_2) lost aerobic respiration reduced
  - pH changes in cytoplasm

  - Proteins denatured
  - Phase transition causes membrane damage
- Slower changes during thawing generally more harmful

Characteristics of Psychrotrophs

- Most important spoilage organisms are aerobes or facultative
- Membrane lipids fluid at low temperature - unsaturated fatty acids
- Extracellular polysaccharides produced - cryoprotectants?
- Enzymes remain active at low temperature
- Enzymes relatively heat-labile

High-Temperature Processes Types of Processes

- Pasteurization destroy pathogens; destroy or reduce spoilage organisms low temp-long hold: 63°C, 30 min. high temp-short time: 72°C, 15 sec.
  - Kill Mycobacterium tuberculosis and Coxiella burnetii
  - Thermodurics and thermophiles survive
- Sterilization destroy all viable organisms commercial sterility: only significant organisms destroyed
- Ultra-high temperatures 140-150°C for a few seconds gives commercial sterility continuous process + aseptic packaging shelf-stable milk

Factors Affecting Heat Resistance

- Moisture dry cells more resistant water may facilitate protein denaturation from heat
- Fat content: higher resistance in presence of fat (lower aw?)
- Salts: some protect (lower aw?), others sensitize
- Carbohydrates: sugars lower aw and increase resistance
- pH: highest resistance at optimum pH (acid foods need less heat)
- Proteins: protective effect
- Number of organisms: higher numberàhigher resistance
- Age of organisms: log phase most sensitive; stationary resistant
- Growth temperature: resistance increases with increasing temp.
- Inhibitory compounds: less resistance if inhibitors present
- Time and temp: greater killing at higher temp, longer time

Microbial heat Resistances

- Resistance related to optimum growth temperature:

    - psychrophiles < mesophiles < thermophiles
    - function of heat sensitivity of enzymes
- Gram + more resistant than Gram -
- Yeasts and molds (including spores) fairly sensitive
- Sporeformers (and their spores) most resistant
    - Protoplast is dehydrated
    - Concentration of ions, polymers (e.g. Ca-dipicolinate)
    - At high growth temperature spore moisture is reduced

Thermal Destruction Concepts

- Target: bacterial endospores, esp. Clostridium botulinum
- Thermal death time: time to kill a given # of org. at a given temp.
- D value:
    - Decimal reduction time
    - Time to destroy 90% of organisms
    - Reciprocal of slope of survivor curve
    - Compare relative heat resistances by comparing D values
- Z value:
    - Of to decrease D by 90%
    - Reciprocal of slope of D vs. oF
    - Can calculate process times at other temperatures
- F value: equivalent time in min. at 250°F of a heat process
- 12 D concept:
    - Process lethality standard in the canning industry
    - Reduce C. botulinum population by 10-12
    - Reduce probability of survival of spores to 10-12
    - For low-acid foods (pH 4.6 and above)

High-heat Processe

- Conventional canning
    - Nonsterile food placed in nonsterile metal or glass container
    - Heat process in closed container
- Aseptic packaging
    - Sterile food placed in sterile containers, sealed aseptically
    - Food pumped through heat exchanger for rapid heating
    - Packaging materials sterilized by hydrogen peroxide

Characteristics of Thermophiles

- Organisms
  - Growth temps: min. 45°C; opt. 50-60°C; max. >70°C
  - Many free-living types
  - Bacilli and clostridia most important in foods
- Growth
  - Short lag time
  - Short log phase
  - Short generation time
  - Often rapid die-off once stationary phase reached
  - May not survive exposure to lower temperatures
- Enzymes
  - High heat resistance → must maintain stable conformation
  - Presence of substrate may increase heat resistance
  - May have high hydrophobic amino acid content
  - May form intramolecular bonds
  - May complex with ions
- Ribosomes
  - More heat-stable than those of mesophiles
  - Higher G-C content more hydrogen bonding
- Nutrient requirements
  - May need more nutrients at higher temperature
  - Less efficient metabolism?
  - Some enzymes inactivated
- Oxygen demand
  - Faster growth rate requires more oxygen
  - Lower oxygen solubility at high temperature
- Lipids and membranes
  - Higher degree of saturation at higher temperature
  - More hydrophobic bonding with increased saturation
  - Membrane stability may be key factor in heat damage

Drying

Types of Foods

- Dry, desiccated: aw = 0-0.6; <25% moisture
  - Combination of dry conditions + heat required
  - Sun drying: raisins

  - Spray and drum drying: nonfat dry milk powder evaporated milk (40% of water left)
- Freeze-dried: same moisture as dry food
  - Freeze product, then sublime water with vacuum (+ heat)
  - Rule of thumb moisture level: 2
- Intermediate moisture: aw = 0.6-0.85; 15-50% moisture
  - Remove water, or add ingredients to tie it up
  - Traditional foods: dried fruits, jams, fermented sausages
  - Newer foods: cat and dog foods, shelf-stable meat products
  - Processes
    a. Adsorption method
    b. First dry food
    c. Rehumidify to adjust aw
    d. Desorption method
    e. Place food in solution of higher osmotic pressure
    f. Allow equilibration
    g. These reach same aw but not necessarily same moisture
    h. Adsorption method holds less water, is more inhibitory
    i. Add humectants (glycerol, sucrose) to tie up water

Effects on Microorganisms

- Preliminary treatments (blanching, alkali, SO_2) reduce load
- Moisture is reduced below that required for growth
  - Bacteria won't grow (Gram- more sensitive than Gram+)
  - Some osmophilic yeasts may grow
  - Molds are primary spoilage organism
- Drying may not kill (unless heat applied) but can injure cells
  - Damage results from ice crystal formation (in freeze-drying)
  - Stress of high solute concentrations in remaining water
- Preservative effect: dry cells less sensitive to heat, radiation
- Expect sporeformers as contaminants of dried foods
- Pathogens of concern: sporeformers, Staph. aureus
- Freeze-drying is popular preservation method for microbial cultures

Fermentation

What is Fermentation

- Carbohydrates and related compounds oxidized to yield energy
- Final electron acceptors are organic compounds derived from carbos

- Only partial oxidation achieved; not all available energy released
- Redox balance of substrate vs. products
- Chemical change brought about in organic substrate by action of microbial enzymes

Features

- Microorganisms responsible for characteristics of food
- Shelf life is usually longer than that of raw product
- Preservation by acid, ethanol, other antimicrobials
- Aroma and flavor due to fermenting organism
- Vitamin content and digestibility can be increased

Types of Fermentations

- Acid food + free sugar: yeasts make alcohol (wines)
- Less acid + free sugar: bacteria make acids (cheese)
- Little simple sugar: require enzymes to break down complex carbohydrates and proteins
  - Malt (germinating barley) in beer
  - Koji (Aspergillus oryzae culture) in soy products (miso)

Products and Processes

- Glucose→lactic acid (homofermentative)
  - Embden-Meyerhof-Parnas (EMP) pathway to pyruvate
- Glucose→acid + CO_2 + ethanol (heterofermentative)
  - Hexose monophosphate or pentose path to pyruvate
- Glucose→pyruvate→ethanol (yeasts)→acetic acid (Acetobacter)
- Glucose or lactate→ propionate + acetate + CO_2 (propionibacteria)

Important Organisms

- Lactic acid bacteria

Lactobacillus	Carnobacterium
Leuconostoc	Enterococcus
Pediococcus	Lactococcus
Streptococcus	Vagococcus

- Yeasts
  - Saccharomyces sp. (esp. S. cerevisiae)
  - Zygosaccharomyces Candida
  - Molds

- Aspergillus Penicillium Geotrichum Rhizopus

Typical Fermentation Process

- Substrate disappears as cell mass increases
- Primary metabolic products (acids) accumulate during growth
- pH drops if acids produced
- Growth and product formation stop as substrate is depleted
- Secondary products (if any) accumulate after growth stops
- For multiple substrates, easiest to metabolize disappears first
- Microbial succession depends on substrate and acid levels
  - Bacteria, then yeasts, then molds grow
  - Sugar, then other small molecules, then polymers used

Food Products

- From milk:
  - Cheese, yogurt, sour cream, buttermilk
  - Lactic acid bacteria (lactobacilli, streptococci) start ferm.
  - Special organisms for unique characteristics
- Meats:
  - Fermented sausages, hams, fish (Asia)
  - Lactic acid bacteria (lactobacilli, pediococci), molds
- From plants:
  - Lactic fermentation (pickles, olives) by lactic acid bacteria
  - Coffee, cocoa beans fermented by yeasts, bacteriaA
  - Soy products (soy sauce, tempeh, miso) by bacteria, molds
- Beverages:
  - Beer and ale (yeasts make ethanol from cereal carbos)
  - Wines (ethanol fermentation from grapes, other fruits)
  - Distilled spirits (must distill to concentrate ethanol)
  - Vinegar (ethanol oxidized to acetic acid)
- Breads:
  - Sourdough (yeast + lactobacilli)
  - Crackers, raised breads (yeasts)

2

What is Microbiology

Microbiology is the study of microorganisms, which are microscopic, unicellular, and cell-cluster organisms. This includes eukaryotes such as fungi and protists, and prokaryotes. Viruses and prions, though not strictly classed as living organisms, are also studied. Microbiology typically includes the study of the immune system, or Immunology. Generally, immune systems interact with pathogenic microbes; these two disciplines often intersect which is why many colleges offer a paired degree such as "Microbiology and Immunology". Microbiology is a broad term which includes virology, mycology, parasitology, bacteriology and other branches.

A microbiologist is a specialist in microbiology and these other topics. Microbiology is researched actively, and the field is advancing continually. It is estimated only about one per cent of all of the microbe species on Earth have been studied. Although microbes were directly observed over three hundred years ago, the field of microbiology can be said to be in its infancy relative to older biological disciplines such as zoology and botany. Bacteria are absolutely necessary for all life on this planet - for every known ecosystem - including the human ecosystem! Without bacteria, there would be no life, as we call life, on the earth.

However, it is a good thing that most bacteria die-out. Here is why: bacteria are single-cell organisms, that produce more of their kind by cell-division, alone. So, if one begins with a single bacterial cell like E. coli for example, in 20 minutes there will be two, and 20 minutes later, four, etc., E. coli cells.

At this rate, even though most bacteria are several hundred-times smaller than we can see with our naked eye (never seen a clothed eye), in only 43 hours, from that one cell at the beginning, there would be enough

E. coli to occupy the entire volume of the earth (1,090,000,000,000,000,000,000 cubic meters)! In only about two additional hours, these bacteria would weigh as much as the earth - 6,600,000,000,000,000,000,000 tons! Bummer! Luckily for us, most bacterial cells die because of the enormous competition for food, and because of other tiny organisms which produce substances (antibiotics) that kill them - you know, like penicillin, which is made by a particular fungus, the mold - Penicillium). Thank goodness for that one, huh?

Actually, many antibiotics are made by certain bacteria too, and, we get many of our necessary vitamins and nutrients from bacteria by allowing the bacteria to multiply in number, and isolating the things that they make, that we cannot make. For example, amino acid supplements are available ("enriched" bread simply means that the amino acid, lysine, which we absolutely need, but cannot make ourselves, is added to the flour used to make the bread), to provide one additional source which most people will eat. This amino acid is produced by certain bacteria grown in huge vats (can be 20,000 litres at one time - that's about 1,500 gallons!), and purified for our use. Antibiotic production is similarly done.

With the advent of molecular genetics and recombinant DNA technology, bacteria now play a very important role as producers of human substances. Since we have learned how genes function, we are able to introduce a human gene into a bacterium and have the product of the human gene expressed. Consequently, a hormone called erythropoietin, which is absloutely necessary for the proper development of red blood cells (erythrocytes), but very, very, difficult to isolate, is now available in high quantity.

People who do not have kidneys cannot make this hormone; however, because the hormone has been cloned into bacteria, plenty of this hormone can be made, purified, and given to these people. Human insulin can be similarly made. These are only two examples of the many substances now available to treat human disorders because of our understanding of bacteria.

THE HISTORY OF MICROBIOLOGY

A look at the history of microbiology will help you to understand the contributions of those who have come before. This perspective will hopefully give you an appreciation of their efforts and put the body of knowledge we will examine in the context of history. Keep in mind that microbiology is a relatively young science. It was only 130 years ago that it became possible to seriously study microorganisms in the laboratory, with most of our understanding of microbes coming in the last 60 years. The history of microbiology, like all human history, is not a catalogue of linear progress, but is more of an interweaving of the careers of bright individuals and their insights.

Each new discovery relied on previous ones and in turn spawned further enquiry. A web of interdependent concepts evolved over time through the work of scientists in many related disciplines and nations. Often the research of one individual impacted the efforts of another studying a completely different problem. Keep this in mind as you look at this history. Below we present several journeys through this web, mentioning some individuals who were particularly important in the progression. This history reflects our view of important events of the past, but is by no means comprehensive.

We will first look at the development of the techniques for handling microorganisms, since everything else in microbiology depends upon these procedures. Next, we will examine how these techniques helped to settle an old debate, the question of spontaneous generation. Then, we will look at the history of infectious disease. The science of microbiology had its most significant early impact on human health, uncovering the cause of the major killers of the day, and then methods to treat them. As microbiology matured, scientists began to look at what non-pathogenic microbes were doing in the environment and we will look a bit at the history of general microbiology. Finally, the chapter will end with an examination of the events that lead to the understanding of life at the molecular level and the profound impact this has had on microbiology and on society in general.

IMPORTANCE OF MICROSCOPES

It took the microscope to expose their tiny world and that instrument has been linked to microbiology ever since. In 1664, Robert Hooke devised a compound microscope and used it to observe fleas, sponges, bird feathers, plants and molds, among other items. His work was published in *Micrographia* and became a popular and widely read book at the time. Several years later Anton van Leeuwenhoek, a fabric merchant and amateur scientist (or "natural philosopher" as such people called themselves), became very adept at grinding glass lenses to make telescopes and microscopes.

While crude by modern standards, his were a technical marvel for the time, able to magnify samples greater than 200-fold. They also produced clearer images than the compound microscopes of the time. By peering through his microscope, Leeuwenhoek observed tiny organisms or "wee animacules" as he called them.. He spent months looking at every kind of sample he could find and eventually submitted his observations in a letter to the Royal Society of London, causing a sensation. Hooke was asked to confirm the findings of Leeuwenhoek and his affirmative assessment garnered them wide acceptance. Surprisingly the work of these two scientists was not followed up for almost 200 years.

Human societies had neither the technical prowess nor the inclination to develop the science of microorganisms. It was not until the rise of the industrial revolution that governments and people dedicated the financial and physical resources to understand these small inhabitants of our world. With the development of better microscopes in the 19th Century, scientists returned to an examination of microorganisms. After finishing his education, Ferdinand Julius Cohn was able to convince his father to lay down the large sum necessary to purchase a microscope for him, one better than that available at the University in Breslau, then part of Germany.

He used it to carefully examine the world of the microbe and made many observations of eukaryotic microorganisms and bacteria. His landmark papers on the cycling of elements in nature was published in *Ueber Bakterien* in 1872 and a microbial classification scheme including descriptions of *Bacillus* were published in the first volume of a journal he founded, *Beitraege zur Biologie der Planzen.* Cohn's work with microscopes popularized their use in microbiology. This and his other work inspired many other scientists to examine microbes. Cohn's encouragement of Robert Koch, a German physician by training, began the field of medical microbiology.

HUMAN EFFECTED BY MICROBES

"Ancient" diseases continue to be a problem where nutrition and sanitation are poor, and emerging diseases such as Acquired Immuno- deficiency Syndrome (AIDS) are even more dangerous for such populations. The Centres for Disease Control and Prevention (the U.S. government agency charged with protecting the health and safety of people) estimates that about 9 per cent of adults between the ages of 18-49 in Sub-Saharan Africa are infected with HIV. AIDS is only one of a number of new diseases that have emerged.

Many of these diseases have no known cure. Influenza and pneumonia are leading killers of the elderly even in the U. S. and other developed nations. Even the common cold causes illness and misery for almost everyone and drains the productivity of all nations. Many of the new diseases are viral in nature, making them notoriously difficult to treat. Disease due to food-borne pathogens also remains a problem, largely because of consumption of improperly processed or stored foods. Understanding the sources of contamination and developing ways to limit the growth of pathogens in food is the job of food microbiologists.

New infections continually appear. Having an available food source to grow on (humans) inevitably results in a microorganism that will take advantage. Some of these feeders will interfere with our own well being, causing disease. Surprisingly, many diseases that were previously thought to have only behavioural or genetic components have been found to involve microorganisms. The clearest case is that of ulcers, which was long thought to be caused by stress and poor diet. However the causative agent is actually a bacterium, *Helicobacter pylori,* and many ulcers can be cured with appropriate antibiotics.

Work on other non-infectious diseases such as heart disease, stroke and some autoimmune diseases also suggest a microbial component that triggers the illness. Finally, some pathogenic microbes that had been "controlled" through the use of antibiotics are beginning to develop drug resistance and therefore re-emerge as serious threats in the industrialized world as well as developing nations. Tuberculosis is an illness that was on the decline until the middle 80's. It has recently become more of a problem, partly due to drug resistance and partly due to a higher population of immunosuppressed individuals from the AIDS epidemic.

Staphylococcus aureus strains are emerging that are resistant to many of the antibiotics that were previously effective against them. These staph infections are of great concern in hospital settings around the world. Understanding both familiar killers and new pathogens will require an understanding of their biology, and thus an understanding of the field of microbiology.

BENEFITS OF MICROBES

Significant resources have been spent to understand and fight disease-causing microorganisms from the beginning of microbiology. You may be surprised to learn that only a small faction of microbes are involved in disease; many other microbes actually enhance our well-being. The harmless microbes that live in our intestines and on our skin actually help us fight off disease. They actively antagonize other bacteria and take up space, preventing potential pathogens from gaining a foothold on our bodies.

Indeed, like all large organisms, humans have entire communities of microorganisms in their digestive systems that contribute to their overall health. The microbial community in humans not only protects us from disease, but also provides needed vitamins, such as B12. Human health and nutrition also depends on healthy farm animals. Cows, sheep, horses and other ruminant animals utilize their microbial associates to degrade plant material into useful nutrients.

Without these bacteria inside ruminants, growth on plant material would be impossible. In contrast to humans ruminant animals have complex stomachs that harbour large numbers of microorganisms. These microbes degrade the plant stuff eaten by the animal into usable nutrients. Without the assistance of the microbes, ruminant animals would not be able to digest the food they eat. Commercial crops are also central to human prosperity, and much of agriculture depends upon the activities of microbes. For example, an entire group of plants, the legumes, forms a cooperative relationship with certain bacteria.

These bacteria convert nitrogen gas to ammonia for the plant, an important nutrient that is often limiting in the environment. Microbes also serve as small factories, producing valuable products such as cheese, yogurt, beer, wine, organic acids and many other items. In conclusion, while it is less apparent to us, the positive role of microbes in human health is at least as important as the negative impact of pathogens. A picture of nodulated leguminous plants. In this case pea plants. Nodules are visible on the roots of the plants in the left of the picture. The plants on the right were not inoculated with nodulating bacteria.

ENVIRONMENT EFFECTED BY MICROBES

The vast majority of life on this planet is microscopic. These teaming multitudes profoundly influence the make-up and character of the environment in which we live. Presently, we know very little about the microbes that live in the world around us because less than 2 per cent of them can be grown in the laboratory. Understanding which microbes are in each ecological niche and what they are doing there is critical for our understanding of the world. Microbes are the major actors in the synthesis and degradation of all sorts of important molecules in environments.

Cyanobacteria and algae in the oceans are responsible for the majority of photosynthesis on Earth. They are the ultimate source of food for most ocean creatures (including whales) and replenish the world's oxygen supply. Cyanobacteria also use carbon dioxide to synthesize all of their biological molecules and thus remove it from the atmosphere.

Since carbon dioxide is a major greenhouse gas, its removal by cyanobacteria affects the global carbon dioxide balance and may be an important mitigating factor in global warming. In all habitats, microorganisms make nutrients available for the future growth of other living things by degrading dead organisms. Microbes are also essential in treating the large volume of sewage and wastewater produced by metropolitan areas, recycling it into clean water that can be safely discharged into the environment.

Less helpfully (from the view of most humans), termites contain microorganisms in their guts that assist in the digestion of wood, allowing the termites to extract nutrients from what would otherwise be indigestible. Understanding of these systems helps us to manage them responsibly and as we learn more we will become ever more effective stewards. Energy is essential for our industrial society and microbes are important players in its production.

A significant portion of natural gas comes from the past action of methanogens (methane-producing bacteria). Numerous bacteria are also capable of rapidly degrading oil in the presence of air and special precautions have to be taken during the drilling, transport and storage of oil to minimize their impact. In the future, microbes may find utility in the direct production of energy. For example, many landfills and sewage treatment plants capture the methane produced by methanogens to power turbines that produce electricity.

Excess grain, crop waste and animal waste can be used as nutrients for microbes that ferment this biomass into methanol or ethanol. These biofuels are presently added to gasoline and thus decreasing pollution. They may one- day power fuel cells in our cars, causing little pollution and having water as their only emission.

Finally, We are increasingly taking advantage of the versatile appetite of bacteria to clean up environments that we have contaminated with crude oil, polycholrinated biphenyls (PCBs) and many other industrial wastes. This process is termed bioremediation and is a cheap and increasingly effective way of cleaning up pollution.

MICROBES AROUND US

Microorganisms will grow on simple, cheap medium and will often rise to large populations in a matter of 24 hours. It is easy to isolate their genomic material, manipulate it in the test tube and then place it back into the microbe. Due to their large populations it is possible to identify rare events and then, with the use of powerful selective techniques, isolate interesting bacterial cells and study them. These advantages have made it possible to test hypothesis rapidly. Using microbes scientists have expanded our knowledge about life. Below are a few examples.

Microorganisms have been indispensable instruments for unlocking the secrets of life. The molecular basis of heredity and how this is expressed as proteins was described through work on microorganisms. Due to the similarity of life at the molecular level, this understanding has helped us to learn about all organisms, including ourselves. Some prokaryotes are capable of growing under unimaginably harsh conditions and define the extreme limits of where life can exist. Some species have been found growing at near 100 °C in hot springs and well above that temperature near deep-sea ocean vents.

Others make their living at near 0 °C in freshwater lakes that are buried under the ice of Antarctica. The ability of microbes to live under such extreme conditions is forcing scientists to rethink the requirements necessary to support life. Many now believe it is entirely possible that Jupiter's moon Europa may harbour living communities in waters deep below its icy crust. What may the rest of the universe hold? Until recently, while we could study specific types of bacteria, we lacked a cohesive classification system, so that we could not readily predict the properties of one species based on the known properties of others.

Visual appearance, which is the basis for classification of large organisms, simply does not work with many microbes because there are few distinguishing characteristics for comparison between species even under the microscope. However, analysis of their genetic material in the past 20 years has allowed such classification and spawned a revolution in our thinking about the evolution of bacteria and all other species. The emergence of a new system organizing life on Earth into three domains is attributable to this pioneering work with microorganisms. The fruits of basic research on microbes has been used by scientists to understand microbial activity and therefore to shape our modern world.

Human proteins, especially hormones like insulin and human growth factor, are now produced in bacteria using genetic engineering. Our understanding of the immune system was developed using microbes as tools. Microorganisms also play a role in treating disease and keeping people healthy. Many of the drugs available to treat infectious disease originate from bacteria and fungi. One last recent role of microbes in informing us about our world has been the tools they provide for molecular biology. Enzymes purified from bacterial strains are useful as tools to perform many types of analyses. Such analyses allow us to determine the complete genome sequence of almost any organism and manipulate that DNA in useful ways.

We now know the entire sequence of the human genome, with the exception of regions of repetitive DNA, and this will hopefully lead to medical practices and treatments that improve health. We also know the entire genome sequence of many important pathogens. Analysis of this data will eventually lead to an understanding of the function of critical enzymes in these microbes and the development of tailor-made drugs to stop them. The tools of molecular biology will also affect agriculture.

For example, we now know the complete genome sequence of the plant *Arabidopsis* (a close relative of broccoli and cauliflower). This opens a new avenue to a better understanding of all plants and hopefully improvements in important crops. Microbes have a profound impact on every facet of human life and everything around us. Pathogens harm us, yet other microbes protect us. Some microbes are pivotal in the growth of food crops, but others can kill the plants or spoil the produce.

Bacteria and fungi eliminate the wastes produced in the environment, but also degrade things we would rather preserve. Clearly they effect many things we find important as humans. In the remainder of this chapter we take a look at how scientists came to be interested in microbes and follow a few important developments in the history of microbiology.

SPECTRUM OF MICROBIOLOGY

Like all other living things, microorganisms are placed into a system of classification. Classification highlights characteristics that are common among certain groups while providing order to the variety of living things. The science of classification is known as taxonomy, and taxon is an alternative expression for a classification category. Taxonomy displays the unity and diversity among living things, including microorganisms. Among the first taxonomists was Carolus Linnaeus. In the 1750s and 1760s, Linnaeus classified all known plants and animals of that period and set down the rules for nomenclature.

CLASSIFICATION SCHEMES

The fundamental rank of the classification as set down by Linnaeus is the species. For organisms such as animals and plants, a species is defined as a population of individuals that breed among themselves. For microorganisms, a species is defined as a group of organisms that are 70 per cent similar from a biochemical standpoint.

In the classification scheme, various species are grouped together to form a genus. Among the bacteria, for example, the species *Shigella boydii* and *Shigella flexneri* are in the genus *Shigella* because the organisms are at least 70 per cent similar. Various genera are then grouped as a family because of similarities, and various families are placed together in an order. Continuing the classification scheme, a number of orders are grouped as a class, and several classes are categorized in a single phylum or division.

The various phyla or divisions are placed in the broadest classification entry, the kingdom. Numerous criteria are used in establishing a species and in placing species together in broader classification categories. Morphology and structure are considered, as well as cellular features, biochemical properties, and genetic characteristics. In addition, the antibodies that an organism elicits in the human body are a defining property. The nutritional format is considered, as are staining characteristics.

PROKARYOTES AND EUKARYOTES

Because of their characteristics, microorganisms join all other living organisms in two major groups of organisms: prokaryotes and eukaryotes. Bacteria are prokaryotes because of their cellular properties, while other microorganisms such as fungi, protozoa, and unicellular algae are eukaryotes. Viruses are neither prokaryotes nor eukaryotes because of their simplicity and unique characteristics.

INDUSTRIAL MICROBIOLOGY/BIOTECHNOLOGY?

Industrial microbiology or microbial biotechnology is the application of scientific and engineering principles to the processing of materials by microorganisms (such as bacteria, fungi, algae, protozoa and viruses) or plant and animal cells to create useful products or processes. The microorganisms utilized may be natural isolates, laboratory selected mutants or microbes that have been genetically engineered using recombinant DNA methods. The terms "industrial microbiology" and "biotechnology" are often one and the same.

Areas of industrial microbiology include quality assurance for the food, pharmaceutical, and chemical industries. Industrial microbiologists may also be responsible for air and plant contamination, health of animals used in testing products, and discovery of new organisms and pathways. For instance, most antibiotics come from microbial fermentations involving a group of organisms called actinomycetes. Other organisms such as yeasts are used in baking, in the production of alcohol for beverages, and in fuel production (gasohol). Additional groups of microorganisms form products that range from organic acids to enzymes used to create various sugars, amino acids, and detergents. For example, the sweetener aspartame is derived from amino acids produced by microorganisms.

INDUSTRIAL APPLICATION OF MICROBES

Microbes have been used to produce products for thousands of years. Even in ancient times, vinegar was made by filtering alcohol through wood shavings, allowing microbes growing on the surfaces of the wood pieces to convert alcohol to vinegar. Likewise, the production of wine and beer uses another microbe — yeast — to convert sugars to alcohol. Even though people did not know for a long time that microbes were behind these transformations, it did not stop them from making and selling these products.

Both of these are early examples of biotechnology — the use of microbes for economic or industrial purposes. This field advanced considerably with the many developments in microbiology, such as the invention of microscope. Once scientists learned about the genetics of microbes, and how their cells produce proteins, microbes could also be altered to function in many new, and useful, ways. This sparked the application of biotechnology to many industries, such as agriculture, energy and medicine.

GENETIC ENGINEERING OF MICROBES

Genetic information in organisms is stored in their DNA. This molecule holds instructions for how the organism looks and functions. DNA is broken into sections called genes, each of which contains the template for a single protein molecule. Proteins serve as building blocks for the cell, and also carry out other activities. By studying microbes, scientists learned how to cut pieces out of a DNA molecule, and move them to another part. This changes how the cell looks or acts. Scientists can also take genes from one organism and insert them into the DNA of another. This gives the organism entirely new abilities.

This type of genetic engineering — the altering of an organism's genetic information — has enabled scientists to use microbes as tiny living factories. One example of this is the production of insulin. In humans, the pancreas creates a protein called insulin that regulates glucose — sugar — levels in the blood. People with one type of diabetes cannot produce insulin, so they inject it into their blood throughout the day. To produce cheaper insulin, scientists inserted the human gene for insulin into the DNA of a common intestinal bacterium. This change enabled the bacterium to produce a new product — human insulin.

Food and Agriculture and Microbes

As with the production of vinegar, microbes are used widely in the agricultural and food industries. Bacteria are used in the production of many food products, such as yogurt, many types of cheese and sauerkraut. Farmers also use a bacterium that produces a natural fertilizer. This type of bacterium is normally associated with bean plants, growing in nodules on the roots in a symbiotic — mutually beneficial — relationship. The bacterium converts nitrogen gas in the air to a form that plants can use — like fertilizer. By adding bacteria to the soil, farmers can increase the productivity of the plants.

Genetic engineering can also be used to produce plants with new abilities, such as enhanced resistance to pesticides, or increased nutritional content. In this case, microbes are used to insert new genes into the DNA of the plants. This results in genetically modified — GM — foods. Humans have long modified the genetics of agricultural plants and animals by breeding them to enhance specific traits. Genetical engineering, however, allows scientists to add genes that exist in totally unrelated organisms.

Energy and Microbes

During vinegar production with wood chips, bacteria grow on the surface of the wood, forming what is called a biofilm. Bacteria attached to a surface like this can produce many compounds, as well as block the flow of a fluid. The latter behaviour has been used to increase the amount of oil extracted from an oil field. Bacteria growing in the wells block areas that are more open. When water is then pumped into the ground, the biofilms drive the water into other areas that still contain oil. This then forces the oil to the surface.

Microbes can also be used to create fuels directly. Certain bacteria ferment glycerol to form ethanol, a biofuel that can be used in automobiles. The glycerol is a byproduct of biodiesel production, but it is more valuable if converted to fuel. With genetic engineering, microbes can also be altered to produce fuels that they don't usually make. One company has modified the DNA of yeast to create biofuel from sugarcane feedstock. The challenge to all of these methods is creating a process that produces fuels more easily — and cheaply — than conventional methods.

Crime and Security and Microbes

Certain types of bacteria thrive in high temperatures. These extremophiles — organisms that prefer extreme conditions — have cell components designed to withstand heat. One of these is a bacterium, Thermus aquaticus, that lives in hot springs and near thermal vents. It contains an enzyme that is involved in the copying of DNA inside the cell. This type of enzyme occurs in other organisms, but the one from T. aquaticus can withstand higher temperatures. Scientists use this enzyme to multiply very small amounts of DNA, such as from samples found at crime scenes. Other techniques are used to identify disease-causing microbes released by terrorists. The microbes can be identified from their DNA. These tests are extremely sensitive, and can find the DNA equivalent of a drop of water in a swimming pool. The U.S. Postal used microbe-detection techniques after letters contaminated with a dangerous microbe — anthrax — were sent through the mail. The tests identified the microbe as coming from the same source, meaning that a single person sent all of the letters.

Medical Application of Microbes

In addition to vaccines and antibiotics, microbes have been essential for many important contributions to medicine. Like diabetes, many diseases can be treated with compounds derived from microbes: cystic fibrosis, cancer, growth hormone deficiency and hepatitis B. In addition, genetic methods that were first developed in microbes now allow scientists to study genetic diseases in humans. This has resulted in the ability to test fetuses for genetic diseases.

There have also been research studies of gene therapy in humans. This technique uses a microbe — often a virus — to insert new genes into cells. In theory, this could correct a condition caused by a genetic disease. Microbial genetics has also led to the ability to determine the sequence of DNA more rapidly, like reading a book. With this information, scientists can look for genes in individuals that cause — or contribute to — diseases.

FACTORS AFFECTING MICROBIAL GROWTH

What do Organisms need for Growth

- Nutrients
  - Water
  - Carbon, nitrogen
  - Organic growth factors
  - Minerals
- Favorable environment
  - Temperature
  - pH
  - Gas mixture
  - No inhibitors
- Time

What Parameters of Foods Affect Microbial Growth

- Intrinsic parameters
  - pH

  - Moisture content
  - Oxidation-reduction potential
  - Nutrient content
  - Antimicrobial constituents
  - Biological structures
- Extrinsic parameters
  - Temperature of storage
  - Relative humidity
  - Gases in the environment

Some General Principles to keep in Mind

- Given good conditions, bacteria grow fastest, then yeast, then mold
- As you move from optimum to stress condition, bacteria give in first,then yeasts, then molds
- Pathogens frequently tolerate stress poorly
- Effects of stress are additive or even synergistic intrinsic parameters

pH

- Measure of hydrogen ion concentration (–log[H+])
- Most organisms grow best around pH 7
- Bacteria have narrowest pH range, then yeasts, then molds
- Pathogens tend to have narrowest pH range
- Organic acids are more inhibitory than inorganic acids
  - Why? →they can penetrate cells more easily
- Foods usually in acid → neutral pH range:
  - Fruits → vegetables → meats, milk
- Buffer capacity reflects resistance to pH change, rate of pH change
- Effects of acids on organisms
  - Energy required to maintain cell's internal pH
  - Enzyme activity affected
  - Proteins, DNA, other molecules denatured
  - longer lag, less rapid growth Water Activity (aw)
- Measure of water available to the organism
- Defined as p/p0 where p = vapor pressure
- Relative humidity (RH) = 100aw
- Aw of fresh foods usually > 0.99

- Bacteria most restricted, then yeasts, then molds
- Spoilage organisms require aw > 0.91; pathogens >0.95
- Various solutes tie up water
  - Degree of inhibition: salt > sugar > glycerol
  - Organisms tolerant of low aw:
  - Halophilic bacteria osmophilic yeasts
  - Xerophilic molds
- Response to low aw
  - Organism concentrates osmoregulators
  - Salts sugars
  - Amino acids polymers
- Effects of low aw
  - Longer lag, slower growth
  - Impaired transport
  - Loss of membrane fluidity

Oxidation/Reduction Potential (Eh)

- Oxidation = loss of electrons; reduction = gain of electrons
- O/R potential = ease with which electrons are lost or gained
- Potential also reflects availability of free oxygen
  - Low potential = anaerobic; high potential = aerobic
- Organisms classified as:
  - Aerobic facultative
  - Anaerobic microaerophilic
- Ranges for microorganisms:
  - Molds are usually aerobic
  - Yeasts usually aerobic or facultative
  - Bacteria from strict anaerobes to strict aerobes
- O/R potential determined by:
  - Characteristic O/R of food
  - Resistance to change (poising capacity)
  - Oxygen tension and access of atmosphere
  - reducing agents in foods: -SH groups, ascorbic acid, sugarspH (low pH high Eh)

Nutrient Content

- Required by all organisms: water, carbon, nitrogen, minerals
- Organic growth factors needed to varying degrees:
  - Gram + > Gram → yeasts > molds
- How easily are energy sources metabolized?
  - Sugar > alcohol > amino acids > complex molecules
- How easily are nitrogen sources metabolized?
  - Amino acids > proteins
  - B vitamins required by many bacteria

Antimicrobial Constituents

- Protective mechanism for living plant/animal?
- Numerous Examples
  - Apices: Essential oils (eugenol in cloves)
  - Milk: Lactoferrin, conglutinin, lactoperoxidase
  - Eggs: lysozyme
  - Fruits: organic acids, hydrocinnamic acid derivatives

Biological Structures

- Protective structures (may be primary defense)
- Skin of fruits (tough protection for seeds)
- Shells of eggs, nuts
- Extrinsic parameters

Temperature

- Growth rate vs. temperature:
  - Rate is highest at optimum temperature
  - Growth slows but may not stop at freezing point
  - rate falls sharply at temperatures above optimum
- Survival vs. temperature:
  - Assume organisms survive freezing
  - Irganisms killed by exposure to high temperatures
  - Temperature range for best growth: 15 - 45°C
- Grosupings based on preferred temperature range:
  - Psychrophiles vs. psychrotrophs
  - Mesophiles

  - Thermophiles vs. thermoduric
- Storage temperature is key factor in stability of perishable foods relative Humidity (RH)
- Should maintain internal aw without adding or losing moisture
- Temperature usually inversely proportional to RH
- Retard surface spoilage by low RH

Gases in the Environment

- Controlled or modified atmosphere storage: increased CO_2
- Retard fungal rot in fruits
  - Decrease pH of meats as carbonic acid forms
- Gas mixtures studied to extend shelf life of meats
  - Floral shift from Gram - to Gram + (more resistant)
  - Shift from aerobes to facultative or anaerobic species
  - will competition among organisms be changed?
  - will pathogens survive where spoilage organisms don't?

3

Alteration Procedure and Part of DNA in Food Microbiology

FOOD PRODUCTS

The first food products containing genetically modified material are now available on supermarket shelves throughout the world. Much public debate has arisen concerning the safety of such products and indeed the need for genetically modified foodstuffs in the well-stocked larders of the Western World.

However, because genetic engineering offers such technical advantages to the food industry in the mass production of cheap processed food of predictable consistency and quality, there will be increased commercial pressure to broaden the range of genetically modified foodstuffs available in the marketplace.

Similarly, as our confidence in the safety of this new technology grows and our ability to apply genetic engineering to products which offer real benefits to the consumer (*e.g.* safer, more wholesome food) and in the development of 'functional' foods which may well play an important role in human health and disease prevention, consumer acceptance and demand for such products may grow.

However, consumer confidence will rely on rigourous assessment of the various potential risks involved and appropriate safety testing. Microorganisms, particularly lactic acid bacteria, have been used in the production of fermented foods for millennia.

Recent advances in genetic engineering allow, for the first time, accurate identification of microorganisms traditionally used in food fermentation and the design of novel strains with improved characteristics. Age-old production problems such as instability of industrially important traits and failure of starter cultures due to bacteriophage attack may now be tackled at the molecular level. It has long been proposed that certain lactic acid bacteria or 'probiotics' contribute greatly to intestinal health and well-being. Similarly, genetically modified lactic acid bacteria have also been proposed for use as oral vaccines.

Recent advances in molecular microbial ecology allow the scientific basis of such claims to be determined and genetic engineering will enable the design of probiotic bacteria with specific health-promoting properties. No food products containing live genetically modified microorganisms (GMMs) are available in the marketplace at present.

However, commercial pressure on the food industry and the consumer benefits promised by probiotic strains designed with scientifically proven health-promoting capabilities will encourage their use in fermented foods in the near future. Clearly the biosafety of such products must be rigourously investigated before they become commercially available, particularly in view of the public back-lash towards genetically modified plant material used in the food products in Europe.

Of particular concern when considering the release of live genetically modified microorganisms in food is the possibility of recombinant DNA transfer from genetically modified organisms (GMOs) to members of the human gut microflora. Transfer of DNA between bacteria occurs naturally in the environment and offers prokaryotes a unique means of evolution and adaptation in response to changing environmental conditions.

Interest in DNA transfer between bacteria in the mammalian gastrointestinal tract stems from three main areas:

- The possibility of DNA transfers from GMOs which may be ingested in food to members of the human gut microflora.
- The spread of antibiotic resistance amongst bacteria as a result of gene transfer.
- The emergence of novel human pathogens as a result of transfer of virulence factors or antibiotic resistance determinants between bacteria.

The focus of this review is to discuss the ability of bacteria to undergo DNA transfer in the human gastrointestinal tract with particular reference to the risks posed by genetically modified microorganisms, which may be used in the production of fermented foods such as bread, beer, cheese and yoghurt. We will look at the existing procedures available to monitor DNA transfer in the human gut microflora and investigate some of the factors governing frequencies of such transfer events. These studies will provide information relevant to our understanding of the possibility of the transfer of marker genes used in the construction of genetically modified crops to the bacteria present in the mammalian gastrointestinal tract. First let us look at the mechanisms of DNA transfer available to bacteria in natural environments. Transformation is the process by which a naked piece of DNA from the environment binds to the surface of a competent bacterial cell and is taken up by the bacterium.

DNA may then be incorporated into the host genome, depending on the recombinational abilities of the host and the 'foreign' DNA. The ability to translocate DNA across the cell boundary is called competence. Competence is a specific physiological state of a bacterial cell (genetically encoded in some cases, *e.g.* Bacillus subtilis), which occurs transiently and is restricted to certain stages of the growth cycle. Natural competence has been observed in bacteria from a variety of genera, including Haemophilus, Neisseria, Streptococcus, and Bacillus, Acinetobacter and Pseudomonas spp. and Helicobacter pylori.

Recalcitrant species, such as Escherichia coli, may be rendered competent using a variety of chemical, enzymatic and physical procedures, *e.g.* CaCl2 treatment and electroporation. Such physiochemical conditions may sometimes be prevalent in the local environment of recalcitrant bacteria. For example, Ca2+ concentrations in drinking water sometimes approach the levels used *in vitro* to induce a state of competence in E. coli.

Competence is not the only bacterial-encoded parameter shown to play a role in natural transformation. In some cases, requirements for specific lengths of DNA, DNA states (double or single stranded) and the presence of specific DNA sequences have been observed. Competent Bacillus and Streptococcus spp. may take up any piece of DNA but only homologous DNA will be maintained in the bacterial genome.

Haemophilus spp., on the other hand, requires the presence of specific 11-bp sequences before DNA uptake occurs. These 11-bp recognition sequences occur at a number of locations on the Haemophilus genome. It has also been reported that a diffusible factor may play a role in the induction of competence in Streptococcus spp. Here, induction of competence was found to be dependent on cell density and the pH of the surrounding environment.

Factors Affecting Transformation in Natural Environments

The presence of naked DNA has been demonstrated in a number of natural environments. Naked DNA may arise in a given habitat via a number of routes. Cell lysis as a result of cell death or the activities of bacteriophage will release bacterial DNA and DNA may be released from actively growing bacteria during certain stages of the growth cycle.

The uptake of DNA from the environment will not only depend on a state of competence in the bacterium but also on the persistence of naked DNA in a given environment. In this respect, different microhabitats will vary greatly in their abilities to either protect naked DNA by the presence of favourable salt concentrations or absorption onto solid supports (*e.g.* the surface of soil particles or food particles in the gut) or to degrade DNA, for example, by active nucleases present in the microenvironment. Despite the longevity of DNA in the soil and the presence of bacteria with known competence for transformation, gene transfer from plant to soil bacteria appears to be an extremely rare event: for example, transfer of an ampicillin resistance gene from a transgenic potato line to the plant pathogenic bacterium Erwinia chrysanthemi was at a calculated frequency of 2 × 10-17 and from transgenic plants to the soil bacterium Acinetobacter calcoacetinus at a frequency lower than 10-13.

However, the uptake and integration of transgenic plant DNA via natural transformation has been shown for Acinetobacter sp., strain BD413 by *in vitro* marker rescue using DNA from various transgenic plants containing the bacterial kanamycin-resistance gene nptII. Naked DNA enters the human gastrointestinal tract from a number of sources. These include ingested food and foreign microorganisms as well as members of the human gut microflora.

Very little is known about the ability of naked DNA to persist in the gut and evade the activities of mammalian and bacterial nucleases. Factors which may affect the persistence of naked DNA, and thus the incidence of transformation in the human gastrointestinal tract, include pH, salt concentrations, cell densities, local nuclease activities and protection afforded by absorption onto surfaces such as food particles or mucosal surfaces.

Such factors would act at the level of the microhabitat and as such would vary greatly even within specific regions of the gut. It has been shown that the CRY1A protein and DNA from genetically modified maize are digested by simulated gastric juices *in vitro.*

However using an *in vitro* model of the intestinal tract, van der Vossen *et al.* showed that 6 per cent of transgenic tomato DNA survived the stomach and small intestine and concluded that the presence of raw mashed tomato helped to preserve the DNA. Free chromosomal DNA of Bacillus subtilis persists for weeks in milk and dairy produce. Such bacteria develop natural competence and can be transformed with free chromosomal or plasmid DNA in such produce. Schubbert *et al.* showed that the wall of the gastrointestinal tract is exposed to a variety of DNA fragments and remains exposed to DNA fragments of dietary origin for hours after ingestion of the food.

The authors found that upon feeding mice 50 μg M13mp18 DNA, approximately 95 per cent of ingested DNA was lost during passage through the stomach. However, phage DNA could be detected by PCR and fluorescent in situ hybridization in peripheral leukocytes, spleen and liver cells as well as the contents of mouse small intestine, caecum, large intestine and faeces. Ingested phage DNA was detected for up to 18 hours after ingestion in caecal contents, for up to 8 hours in DNA from the peripheral blood cells and for up to 24 hours but not 48 hours in DNA from spleen and liver. Mercer *et al.* monitored the survival of recombinant plasmid DNA (pVACMC1) in fresh human saliva. The fraction of naked DNA remaining amplifiable in saliva ranged from between 40 and 65 per cent after 10 minutes and 6-25 per cent after 60 minutes. Amplifiable plasmid DNA was still present after 24 hours incubation in fresh saliva.

The authors also found that plasmid DNA, which had been exposed to degradation by saliva, was capable of transforming naturally competent Streptococcus gordonii DL1 in filtered saliva. Transformation activity decreased rapidly with the extent of plasmid DNA degradation. Such studies suggest that DNA released from food or bacteria ingested with food may undergo transformation in not only the oral cavity but other regions of the human gastrointestinal tract. Clearly, much more research is needed on the ability of naked DNA to persist in the lower regions of the human gastrointestinal tract and the extent to which transformation contributes to DNA transfer in the human gut microflora.

CONTROL IN FOOD PROCESSING

The maintenance of a well functioning Quality Assurance (QA) programme is essential if a consistent product is to result which meets all required standards. Such a programme should be based on Hazard Analysis and Quality Analysis Critical Control Point (HACCP and QACCP) systems. HACCP and QACCP are more proactive than traditional approaches to QA/QC activities. The establishment of such programmes is the responsibility of QA personnel but the execution of it involves everyone in the company.

To avoid ambiguity regarding responsibility for any QA function, it is important to assign specific HACCP/QACCP accountabilities to responsible persons and groups. A QA programme must consider all activities impacting upon product quality, from raw materials and ingredients used to product handling through distribution channels all the way to the final consumer.

In respect of this, Wilson has outlined the following required components of a QA system:

- Raw material control – standard specifications must be adopted for all ingredients which must then be inspected to ensure conformity;
- Process control – all chemical, physical and microbiological hazards as well as quality factors must be identified, critical control points (CCP) must be established, monitored and a record made of any action taken;
- Finished product control–this requires that the finished product be unadulterated, properly labelled and that the integrity of the finished be protected from the environment.

MECHANISM OF TRANSDUCTION OF FOOD

Transduction is the mechanism by which DNA may be transferred between bacteria by bacteriophage. In essence, the bacteriophage acts as microbial couriers, picking up DNA from one bacterial chromosome and delivering the heterologous DNA to another bacterial chromosome. Where degradation of naked DNA is one of the chief factors limiting DNA transfer by bacterial transformation in natural environments, bacteriophage protects DNA in natural environments via their proteinaceous capsid. Bacteriophages, on the whole, contribute to gene transfer between bacteria by two main mechanisms, namely specialized and generalized transduction.

Generalized Transduction

Generalized transduction on the other hand, occurs when host DNA is packaged into phage particles instead of the phage genome. This mechanism of transduction has been observed in the lytic phage, P1 and P2 of the enterobacteraceae, for example. Upon transduction to a novel recipient, the transferred DNA may either be incorporated into the host chromosome via recombination or where it possesses the means of self-replication, it may replicate autosomally.

Specialized Transduction

Specialized transduction involves the incorporation of a lysogenic phage into the bacterial chromosome. Lysogeny is favoured by conditions of environmental stress such as the limitation of nutrients and probably aids the survival of the phage and host bacterium. Upon excision from the chromosome, elements of host DNA adjacent to the prophage DNA may become excised along with the phage DNA.

The host DNA may then become incorporated into the bacteriophage genome and packaged along with the phage DNA. Thus, when infection of a recipient bacterium occurs the heterologous DNA may become integrated into the recipient's chromosome during lysogeny.

Factors Limiting DNA Transfer by Transduction

Transduction is greatly dependent on the host range of the transducing bacteriophage, which is generally narrow, since bacteriophage infection is dependent on the phage recognizing specific receptor sites on the bacterial cell surface. Some bacteriophage that are able to mediate transduction between different species of bacteria have been described, *e.g.* P1 and Mu.

The frequency of transduction in nature may be much higher than previously recognized, since the number of bacteriophage particles in many environments appears to be much higher than first thought. Whether the transferred DNA is maintained in the new host is greatly dependent on the ability of the bacteriophage to insert itself and the heterologous DNA into the host genome and evade the bacterial restriction modification system. Bacterial restriction enzymes may recognize specific sequences on heterologous or phage DNA. It has been proposed that restriction modification systems may reduce infection of unmodified phage DNA by 2-3 orders of magnitude. The amount of DNA that may be transferred by transduction is also limited, being about equivalent in size to the phage genome itself.

Transduction in the Human Gastrointestinal tract

Despite the fact that the bacteriophage are ubiquitous members of natural microbial ecosystems, little is known about their distribution or activity in the human gut microflora. To date no reports of transduction in the human gastrointestinal tract have been presented. However, where sufficient effort has been made to isolate bacteriophage from natural environments and examine their activity in situ, they have been found to play a significant role in microbial ecology.

The fact that about 90 per cent of all bacteriophage isolated from natural environments are temperate suggests that transduction may be more important in microbial genetic plasticity than previously appreciated. Thus specifically designed studies involving *in vitro* or *in vivo* models of the human gastrointestinal microflora may well provide evidence of transduction in this ecosystem and elucidate some of the ecological factors governing transduction.

Bacteriophage specific for major groups of bacteria present in the gut microflora, *e.g.* Bacteroides spp., bifido-bacteria, lactobacilli, methanogens, clostridia, enterobacteriaceae, streptococci and staphylococci have been identified. However, transduction has not been observed in many of these bacterial groups *e.g.* Bacteroides spp. or Clostridium spp.

Bacteriophage of the lactic acid bacteria play a major role in the dairy industry both as destructive agents, causing the failure of starter cultures in cheese and yoghurt production, and as valuable genetic tools in the development of genetically altered industrial strains. The variety of bacteriophage associated with the lactobacilli suggests that transduction may be a significant means of gene transfer within these genera.

Tohyama *et al.* demonstrated that the Lactobacillus salivarius temperate phage PLS-1 mediates generalized transduction of auxotrophic markers (lysine, proline and serine) and lactose metabolism at frequencies of 10-7 to 10-8 transductants per CFU *in vitro*. Later, Raya *et al.* showed that phage ADH replicates in a lytic cycle, establishes lysogeny, confers superinfection immunity on the host and mediates plasmid DNA transduction in L. acidophilus ADH. It was also observed that plaque formation on cell lawns of L. acidophilus NCK102 was pH dependent, with the optimal pH for plaque formation being pH 5.5. Transfer of antibiotic resistance determinants have been reported between strains of Desulfovibrio desulfricans via phage Dd1 mediated transduction at frequencies of 10-5 to 10-6 transductants per recipient.

Transduction has also been observed in methanogens, although not strains found in the human gut microflora. Despite demonstrations that transduction does occur under labouratory conditions, little is known about the significance of transduction-mediated gene transfer in the natural and animal-associated environments. Important questions remain unanswered regarding the prevalence and survival of bacteriophage in different environments, the limitations imposed on transduction by the host specificity of the transducing bacteriophage and the frequencies at which transduction occurs during phage replication.

GENETIC ELEMENTS IN BACTERIAL CONJUGATION

Plasmid DNA, extrachromosomal, self-replicating genetic elements, may mediate DNA transfer between bacteria via conjugation. Conjugation involves cell-to-cell contact between a donor (plasmid-bearing) and recipient (plasmid-free) cell. Not all plasmids are conjugative but many conjugative plasmids carry environmentally important traits such as antibiotic resistance determinants, virulence factors, novel degradative pathways.

Different bacteria or groups of bacteria display different mechanisms of conjugation, a comprehensive discussion of which is beyond the scope of this review. However, a short, generalized description of some of the conjugative mechanisms employed by different groups of bacteria may serve to highlight both the mechanistic diversity of bacterial conjugation and that DNA transfer via conjugation-like mechanisms has been observed in bacteria from a wide phylogenetic background.

Conjugation in Gram-positive Bacteria

Conjugation of plasmids in Gram-positive bacteria differs considerably from those in Gram-negative bacteria. Two main mechanisms have been described. Certain plasmids found in Enterococcus spp. and Lactococcus spp. have been shown to initiate conjugation in response to a small peptide signal released extracellularly by plasmid-free recipient cells. These extracellular peptides act as sex pheromones, and lead to clumping of donor and recipient cell to form a conjugative aggregate. This increases cell-to-cell contact between donor and recipient cells and results in the transfer of pheromone-induced plasmids at high frequencies.

Plasmid pAD1 of Enterococcus faecalis is the best characterized example and is transferred at high frequencies in conjugative aggregates formed in response to pheromones released by plasmid-free recipient cells. The second type of conjugation elucidated in Gram-positive bacteria, especially Streptococcus spp. and Staphylococcus spp., depends on the direct contact between donor and recipient cell imposed by growth on a solid surface.

Plasmids employing this form of DNA transfer are not induced to conjugate by sex pheromones and the mechanism of their transfer remains unclear. Such plasmids often display a remarkable broad host range, *i.e.* they can be transferred to a wide spectrum of both Gram-positive and Gram-negative bacteria, including Streptococcus spp., Staphylococcus spp., Clostridium spp., Lactobacillus spp., Listeria spp., Pediococcus spp. and Bacteroides spp. Many of these plasmids encode one or more antibiotic-resistance determinants and are thought to have played a role in the dissemination of antibiotic-resistance determinants among bacteria of clinical importance. The best characterized examples include pAMb1 and pIP501.

Conjugation in Archaea

Little is known about the molecular biology of Archaea because of the difficulties in cultivating such organisms routinely under labouratory conditions and the major differences that exist between Archaea and Bacteria. However, a conjugation-like mechanism of DNA transfer has been observed in the halophilic Archaea Halobacterium volcanii.

Here, a 'cytoplasmic' bridge forms between parental cells along which chromosomal DNA has been shown to transfer. The mechanism of DNA transfer displays characteristics of both bacterial conjugation and eukaryotic cell fusion. However, until we gain a greater understanding of the molecular biology and indeed microbiology of the Archaea, few conclusions can be drawn about the frequencies or mechanisms of DNA transfer in Archaea.

Conjugative Transposon

Transposons, genetic elements borne by microbial genomes and possessing the ability to 'transpose' between different locations on the same genome, have long been known to play a role in microbial genetic plasticity. Some such elements not only possess the machinery required for transposition but are also capable of mediating their own transfer between the genomes of different bacteria via conjugation. Such conjugative transposons have been found in a wide array of both Gram-positive and Gram-negative bacteria.

They are thought to play a significant role in the dissemination of antibiotic- resistance determinants among a wide range of bacteria and the emergence of multiple antibiotic-resistant pathogenic strains. Conjugative transposons generally excise from the bacterial chromosome to form a covalently closed circular double-stranded DNA transposition intermediate. The transposition intermediate can either reinsert itself into the bacterial chromosome or onto a plasmid in the same cell, or it can mediate its own conjugation to a second bacterial cell and integrate into the recipient chromosome. Conjugative transposons have a wide host range and have been shown to transfer between bacteria at high frequencies typically 10-5 to 10-4 per recipient.

Conjugation in Gram-negative Bacteria

In Gram-negative bacteria, cell-to-cell contact between a donor and recipient cell is accomplished via the formation of a pilus (which can be short and rigid or long and flexible) by the donor cell. This pilus, a proteinaceous tube, forms a cytoplasmic bridge between donor and recipient, along which a single-stranded copy of the conjugative plasmid is transferred to the recipient cell. DNA replication then leads to generation of the complementary strand of plasmid DNA in both recipient and donor. Certain plasmids originally found in Gram-negative species display an extremely broad host range.

Such plasmids (*e.g.* the R plasmids) may carry a number of antibiotic-resistance determinants and have been shown to transfer to a variety of bacterial species in different natural environments. Some conjugative plasmids may mediate transfer of chromosomal DNA between donor and recipient cells upon integration into the chromosome via homologous recombination and low-fidelity excision from the chromosome, leading to incorporation of segments of bacterial DNA into the circularized plasmid before transfer. Transfer of Gram-negative plasmids has been observed in many natural environments, including soil, rhizosphere, salt and fresh water and effluent.

Mobilization

Some conjugative plasmids may 'mobilize' plasmids co-resident in a bacterial cell. Mobilization results in the transfer of a non-self transmissible plasmid from one bacterial cell to another and is mediated by the conjugative plasmid or conjugative transposon. Mobilization may occur as a result of the non-conjugative plasmid 'hijacking' the conjugative machinery of the conjugative plasmid or the two plasmids may form a co-integrate plasmid. Plasmids with oriT sequences may be mobilized upon formation of efficient cell-to-cell contact between donor and recipient bacteria mediated by the trans acting products of tra and mob genes borne on self-conjugative plasmids co- resident in the same cell.

Plasmids devoid of oriT sequences may be mobilized by formation of a co-integrate with a self-conjugative plasmid co-resident in the same cell. Such co-integrate formation is characteristic of bacterial transposable elements. The co-integrate once transferred to a recipient strain via conjugation, may then resolve into its constituent plasmids. Mobilization of genetically modified plasmids has been demonstrated from introduced genetically modified donor strains to bacteria present in different environments. Such observations have implications for the release of genetically modified microorganisms into natural environments, even when steps have been taken to limit the transfer potential of recombinant DNA.

Many broad host range plasmids are also mobilizable, which aids their dissemination within microbial consortia as well as the dissemination of antibiotic resistance often borne on such plasmids. Some conjugative plasmids can also mobilize chromosomal DNA between different bacteria. Conjugative transposons have also been shown to mediate mobilization of non-conjugative plasmids or chromosomal genetic elements between bacteria.

EFFECTS ON FOOD PRODUCTION

Assuming no effects of climate change on crop yields and current trends in economic and population growth rates, world cereal production5 is estimated at 3286 million metric tons (mmt) in 2060 (cf. 1795 mmt in 1990). Per capita cereal production in developed countries increases from 690 kg/cap in 1980 to 984 kg/cap in 2060.

In developing countries (excluding China) cereal production increases from 179 to 282 kg/cap. Aggregated world per capita cereal production grows from 327 kg/cap in 1980 to 319 kg/cap in 2060. The declining aggregate trend for the future is caused by the relatively large difference in per capita cereal production in the developed and developing countries and the demographic changes assumed by the model.

Cereal prices are estimated at an index of 121 (1970 = 100) for the year 2060, reversing the trend of falling real cereal prices over the last 100 years. This occurs because the BLS standard reference scenario has two phases of price development.

During 1980 to 2020, while trade barriers and protection are still in place but are being reduced, there are increases in relative prices; price decreases follow when trade barriers are removed. The number of hungry people is estimated at about 640 million or about 6 per cent of total population in 2060 (cf. 530m in 1990, about 10 per cent of total current population).

ADJUSTMENTS IN THE ECONOMIC SYSTEM

The BLS includes the ability to simulate adjustments that the world food system might make to changes of yield (*e.g.*. reallocation of agricultural land use, change in fertilizer use, and application of irrigation water). Simulations of the effects of climate change without such internal adjustments are of theoretical interest only as these would unrealistically imply no economic or behavioural response of producers and consumers.

However, as a measure of distortion of the economic system these hypothetical impacts help to define the adjustments taking place in the system over time. Under these conditions the effects of climate change and increased atmospheric CO2 on crop yields derived from the GCM scenarios imply a 5 per cent to almost 20 per cent reduction in total cereal production. These estimates are changes to production levels projected for 2060 without climate change.

Adjustments within the economic system tend to counteract negative yield impacts as agricultural production shifts to regions of more favourable comparative advantage. The BLS offsets 65-80 per cent of the potential impact on yield in scenarios for impacts below 10 per cent of global cereal production (the GISS and GFDL climate change scenarios). The offset decreases to 60 per cent under a scenario of greater yield reduction (*e.g.*., UKMO).

Adjustment in the Economic System

Changes in cereal production, cereal prices, and people at risk of hunger estimated for the GCM doubled CO2 climate change scenarios (with the direct CO2 effects taken into account). These estimations are based upon dynamic simulations by the BLS that allow the world food system to respond to climate- induced supply shortfalls of cereals and consequently higher commodity prices through increases in production factors (cultivated land, labour, and capital) and inputs such as fertilizer.

The testing of climate change impacts without farm-level adaptation is unrealistic, but is done for the purpose of establishing a baseline with which to compare the effects of farmer response. We can safely assume that at least some farm-level adaptations will be adopted, especially techniques similar to those tested in Adaptation Level 1 that do not imply major changes to current agricultural systems.

Under the GISS scenario (which provides lower temperature increases) cereal production is estimated to decrease by just over 1 per cent, while under the UKMO scenario (with the highest temperature increases) global production is estimated to decrease by more than 7 per cent. The largest negative changes occur in developing regions, which average-9 per cent to-11 per cent, though the extent of decreased production varies greatly by country depending on the projected climate. By contrast, in developed countries production is estimated to increase under all but the UKMO scenario (+11 per cent to-3 per cent). Thus, disparities in crop production between developed and developing countries are estimated to increase.

Price increases resulting from climate-induced decreases in yield are estimated to range between 25 per cent and 150 per cent. In the case of the GISS scenario, the 5.3 per cent reduction in yields of the unadjusted scenario causes a disequilibrium that is resolved via market mechanisms in the adjusted case. This results in a-1.2 per cent consumer response and about a +4 per cent (relative) producer response and leads to 24 per cent higher relative prices for cereals. While this price response seems to be high, cereal prices only account for a modest fraction, perhaps one third or less, of retail food prices. Hence, a 24 per cent increase in world cereal prices does not imply a 24 per cent increase in food prices.

These increases in price are likely to affect the number of people with insufficient resources to purchase adequate amounts of food. The estimated number of hungry people increases approximately 1 per cent for each 2-2.5 per cent increase in prices (depending on climate change scenario). People at risk of hunger increase by 10 per cent to almost 60 per cent in the climate change scenarios tested, resulting in an estimated increase of between 60 and 350 million people in this condition (above the reference case of 640 million) by 2060.

Farmer Adaptation

Globally, both minor and major levels of adaptation help restore world production levels, compared to the climate change scenarios with no adaptation. Averaged global cereal production decreases by up to about 160 mmt (0 per cent to-5 per cent) from the reference case of 3286 mmt with Level 1 adaptations. These involve shifts in farm activities that are not very disruptive to regional agricultural systems. With adaptations implying major changes, global cereal production responses range from an additional 30 mmt to slight increase to a slight decrease of about 80 mmt (+1 per cent to-2.5 per cent).

Level 1 adaptation largely offsets the negative climate change yield effects in developed countries, improving their comparative advantage in world markets. In these regions cereal production increases by 4 per cent to 14 per cent over the reference case. However, developing countries are estimated to benefit little from this level of adaptation (-9 per cent to-12 per cent change in cereal production). More extensive adaptation (Level 2) virtually eliminates global negative cereal yield impacts derived under the GISS and GFDL climate scenarios and reduces impacts under the UKMO scenario to one third.

The effects of climate change, and climate change with both levels of adaptation, on cereal prices in 2060. As a consequence of climate change, world cereal prices are estimated to increase by about 25 per cent to almost 150 per cent. Under Adaptation Level 1, price increases range from 10 per cent to 100 per cent; under Adaptation Level 2, cereal price responses range from a decline of about 5 per cent to an increase of 35 per cent.

As a consequence of climate change and Adaptation Level 1, the number of people at risk from hunger increases by about 40 million to 300 million (6 per cent to 50 per cent) from the reference case of 641 million. With more significant farmer adaptation (Level 2), the number of people at risk from hunger is altered by between-12 million for the GISS scenario and 120 million for the UKMO scenario (-2 per cent and +20 per cent).

These results indicate that, except for the GISS scenario under Adaptation Level 2, the simulated farm-level adaptations did not entirely mitigate the negative effects of climate change on potential risk of hunger, even when economic adjustments, *i.e.*. the production and price responses of the world food system, are taken into account. For each of these alternate future assumptions, a new reference scenario was established with the BLS, and then tested with the GCM climate change scenarios.

Full trade liberalization: Assuming full agricultural trade liberalization and no climate change by 2020 provides for more efficient resource use. This leads to a 3.2 per cent higher value added in agriculture globally and 5.2 per cent higher agricultural GDP in developing countries (excluding China) by 2060 compared to the original reference scenario. This policy change results in almost 20 per cent fewer people at risk from hunger. Global cereal production increases by 70 mmt, with most of the production increases occurring in developing countries. Climate change impacts were then simulated under these new reference conditions. Under the same trade liberalization policies, global impacts due to climate change are slightly reduced, with enhanced gains in production accruing to developed countries. Losses in production are greater in developing countries. Price increases are reduced slightly from what would occur without full trade liberalization, and the number of people at risk from hunger is reduced by about 100 million.

Reduced rate of economic growth: Estimates were also made of impacts under a lower economic growth scenario (10 per cent lower than reference). Lower economic growth results in a tighter supply situation, higher prices, and more people below the hunger threshold. The effect of climate change on these trends is generally to reduce production, increase prices, and increase the number of people at risk from hunger. Developed countries increase cereal production in the GISS and GFDL scenarios even with the projected lower economic growth rates, but developing countries decrease production under all climate change scenarios.

Altered rates of population growth: Lower population growth has a significant effect on cereal production, food prices, and number of hungry people. Simulations based on rates of population growth according to UN Low Estimates result in a world population about 17 per cent lower in year 2060 when compared to UN Mid Estimates as used in the reference run. The corresponding reduction in the developing countries (excluding China) would be about 19.5 per cent, from 7.3 billion to 5.9 billion. The combination of higher GDP/capita (about 10 per cent) and lower world population produces an estimated 40 per cent fewer people in hunger in the year 2060 compared to the reference scenario.

Even under the most adverse of the three climate scenarios (UKMO) the estimated number of hungry is some 10 per cent lower than the estimated reference scenario without any climate change. Increases in world prices in agricultural products, in particular of cereals, under the climate change scenarios employing the low population projection are around 75 per cent of those using the UN Mid Estimate.

The generalized relative effects of different policies regarding trade liberalization, economic growth and population growth on the production of cereals and people at risk of hunger. Alternative development assumptions make little difference with respect to the geopolitical patterns of the relative effects of climate change. In all cases, cereal production decreases, particularly in the developing world, while prices and population at risk from hunger increase due to climate change.

The beneficial effects of trade liberalization and low population growth are of the same or even greater (in the case of population) order of magnitude as the adverse effects of climate change. This suggests that there may be much to be gained from altering the conditions of trade and development as a strategy for addressing the climate change issue. The magnitude of adverse climate impacts are least, however, under the conditions of low population growth. An assumption of low population growth rate minimizes the population at risk from hunger both in the presence and absence of climate change in the BLS simulations.

Climate change induced by increasing greenhouse gases is likely to affect crop yields differently from region to region across the globe. Under the climate change scenarios adopted in this study, the effects on crop yields in mid-and high-latitude regions appear to be less adverse than those in low-latitude regions. However, the more favourable effects on yield in temperate regions depend to a large extent on full realization of the potentially beneficial direct effects of CO2 on crop growth.

Decreases in potential crop yields are likely to be caused by shortening of the crop growing period, decrease in water availability due to higher rates of evapotranspiration, and poor vernalization of temperate cereal crops. When adaptations at the farm level were tested (*e.g.*. change in planting date, switch of crop variety, changes in fertilizer application and irrigation), compensation for the detrimental effects of climate change was more successful in developed countries.

When the economic implications of these changes in crop yields are explored in a world food trade model, the relative ability of the world food system to absorb impacts decreases with the magnitude of the impact. Regional differences in effects remain noticeable: developed countries are expected to be less affected by climate change than developing economies. Dynamic economic adjustments can compensate for lower-impact scenarios such as the GISS and GFDL climate scenarios but not higher-impact ones such as the UKMO scenario. Prices of agricultural products are related to the magnitude of the climate change impact, and incidence of food poverty increases even in the least negative climate change scenario tested.

When the effects of lower future population and economic growth rates and trade liberalization were tested in the food trade model, reduced population growth rates would have the largest effect on minimizing the impact of climate change. Lower economic growth results in tighter food supplies, and consequently would result in higher rates of food poverty.

Full trade liberalization in agriculture, on the other hand, provides for more efficient resource use and would reduce the number of people at risk from hunger by about 100 million (from the reference case of c. 640 million in 2060). However, all of the scenarios of future climate adopted in this study increase the estimates of the number of people at risk from hunger. It should be emphasized that the results reported here are not a forecast of the future.

There are very large uncertainties that preclude this: particularly the lack of information on possible climate change at the regional level, the effects of technological change on agricultural productivity, trends in demand (including population growth), and the wide array of possible adaptations. The adoption of efficient adaptation techniques is far from certain. In developing countries there may be social or technical constraints, and adaptive measures may not necessarily result in sustainable production over long timeframes. The availability of water supplies for irrigation and the costs of adaptation are both critical needs for further research.

Future trace gas emission rates, as well as when the full magnitude of their effects will be realized, are not certain, and only a limited range of GCM climate change scenarios, representing the upper end of the projected warming, was tested. However, it can be argued that the use of scenarios from the higher GCM projections provides perspective on the downside risk of global warming projections. Because of these uncertainties, the study should be considered as an exploratory assessment of the sensitivity of the world food system to a limited number of what is, in effect, a much wider array of possible futures.

Determining how countries, particularly developing countries can and will respond to reduced yields and increased costs of food is a critical research need arising from this study. Will such countries he able to import large amounts of food? From a political and social standpoint, these results show a decrease in food security in developing countries. The study suggests that the worst situation arises from a scenario of severe climate change, low economic growth, and little farm-level adaptation.

In order to minimize possible adverse consequences-production losses, food price increases, and people at risk of hunger-the way forward is to encourage the agricultural sector to continue to develop crop breeding and management programmes for heat and drought conditions (these will be immediately useful in improving productivity in marginal environments today), in combination with measures taken to slow the growth of the human population of the world. The latter step would also be consistent with efforts to slow emissions of greenhouse gases. The source of the problem, and thus the rate and eventual magnitude of global climate change.

TRANSFER OF PLASMID DNA IN RETROTRANSFER

Transfer of plasmid DNA between bacteria is usually unidirectional, *i.e.* from donor to recipient. However, Mergeay *et al.* demonstrated the transfer of genetic markers from a recipient strain to the donor strain during conjugation experiments. This form of plasmid transfer has been called retrotransfer. The mechanism involved has not been fully elucidated, but is thought to involve multiple donor and recipient cells. Retrotransfer has been Demonstrated in a number of bacterial species.

Factors governing DNA Transfer

Plasmid transfer by conjugation is determined by characteristics of the plasmid itself, *e.g.* tra+, mob+ and oriT+ genotypes, and the donor and recipient bacteria involved. Environmental parameters which affect pilus formation or the rate of plasmid curing, such as SDS, organic solvents, proteases, high temperature and acridine orange, will affect the frequency of conjugation. The metabolic character of the host bacterium, *e.g.* growth rate or stress, may also affect the frequency of conjugation.

The frequency of conjugation may be enhanced by the presence of surfaces and bacterial aggregation (*e.g.* mediated by bile salts or sex pheromones). The presence of solid surfaces, *e.g.* soil particles, food particles or walls of mucosa, in an environment may also affect the frequency of plasmid transfer depending on the mode of conjugation employed, by increasing cell-to-cell contact between donor and recipient cells. Substrate availability may also affect conjugation with increased frequencies of conjugation sometimes observed with a plentiful supply of bacterial substrates.

Physiological conditions extant in the environment, such as pH, temperature, substrate availability and bacterial cell density, may all play a role in governing the frequency of conjugation in a plasmid- and bacterium-specific manner. Plasmid maintenance by bacteria exacts a metabolic cost on the cell, *i.e.* the energy needed for the plasmid to replicate and express plasmid encoded proteins.

This 'metabolic burden' may be counteracted by adaptational advantages endowed by plasmid maintenance. Plasmids often encode environmentally important traits such as resistance to antibiotics and bacteriophage, unusual metabolic degradative pathways, virulence factors and genes encoding anti-bacterial compounds. Such traits may enable a plasmid-bearing cell to out-compete a plasmid-free cell in response to altering environmental conditions, *e.g.* presence of antibiotics, unusual substrates or bacteriophage.

Similarly, environmental factors affecting the competitiveness or growth rate of a bacterial cell will affect the ability of the cell to both undergo conjugation and maintain newly acquired plasmids. Properties of the plasmids themselves may limit the extent to which conjugation occurs in the environment. For example plasmid incompatibility, host range, surface exclusion, evasion of host restriction modification systems and segregational stability may all play a role in the extent to which a particular plasmid may spread and persist in natural microbial communities.

Thus it appears that the frequency of plasmid transfer in a given environment will be a function of the interaction between environmental factors and the bacterial populations involved, characteristics of the plasmid itself and the mode of conjugation employed by the plasmid. Assessment of the Safety of BD Foods Current safety assessment methodologies are focused primarily on the evaluation of the toxicity of single chemicals. Food is a complex mixture of many chemicals. Using animal models, the evaluation of most aspects of the safety of single components of the diet, such as a Bt toxin, is possible using widely accepted protocols. Future projects may involve more complicated manipulations of plant chemistry.

In this case, safety testing will be more challenging. Whole foods cannot be tested with the high dose strategy currently used for single chemicals to increase the sensitivity in detecting toxic endpoints. Also, the question of potential deleterious interactions between new or enhanced levels of known toxic agents in BD foods will undoubtedly be raised. The safety testing of multiple combinations of chemicals remains a difficult proposition for toxicologists. In view of these challenges, there is a clear need for the development of effective protocols to allow the assessment of the safety of whole foods.

The responsibility of toxicologists is to assess whether foods derived through biotechnology are at least as safe as their conventional counterparts and to ascertain that any levels of additional risk are clearly defined. In achieving this goal, it is important to recognize that it is the food product itself, rather than the process through which it is made, that should be the focus of attention. In assessing safety, the use of the substantial equivalency concept provides guidance as to the nature of any new hazards.

Scientific analysis indicates that the process of BD food production is unlikely to lead to hazards of a different nature from those already familiar to toxicologists. The safety of current BD foods, compared with their conventional counterparts, can be assessed with reasonable certainty using established and accepted methods of analytical, nutritional, and toxicological research.

A significant limitation may occur in the future if transgenic technology results in more substantial and complex changes in a foodstuff. Methods have not yet been developed by which whole foods (as compared with single chemical components) can be fully evaluated for safety. Progress also needs to be made in developing definitive methods for the identification and characterization of protein allergens, and this is currently a major focus of research.

Improved methods of profiling plant and microbial metabolities, proteins, and gene expression may be helpful in detecting unexpected changes in BD organisms and in establishing substantial equivalence. The level of safety of current BD foods to consumers appears to be equivalent to that of traditional foods. Verified records of adverse health effects are absent, although the current passive reporting system would probably not detect minor or rare adverse effects, nor can it detect a moderate increase in common effects such as diarrhea. However, this is no guarantee that all future genetic modifications will have such apparently benign and predictable results. A continuing evolution of toxicological methodologies and regulatory strategies will be necessary to ensure that this level of safety is maintained. The term *biotechnology* refers to the use of scientific techniques, including genetic engineering, to improve or modify plants, animals, and microorganisms.

In its most basic forms, biotechnology has been in use for millennia. For example, Middle Easterners who domesticated and bred deer, antelope, and sheep as early as 18,000 B.C.E.; Egyptians who made wine in 4000 B.C.E.; and Louis Pasteur, who developed pasteurization in 1861, all used biotechnology. In recent years, however, food biotechnology has become synonymous with the terms *genetically engineered foods* and *genetically modified organism* (GMO).

Traditional biotechnology uses techniques such as crossbreeding, fermentation, and enzymatic treatments to produce desired changes in plants, animals, and foods. Crossbreeding plants or animals involves the selective passage of desirable genes from one generation to another. *Microbial* fermentation is used in making wine and other alcoholic beverages, yogurt, and many cheeses and breads. Using enzymes as food additives is another traditional form of biotechnology. For example, papain, an enzyme obtained from papaya fruit, is used to tenderize meat and clarify beverages.

GENETIC ENGINEERING

The DNA contained in genes determines inherited characteristics. Modifying DNA to remove, add, or alter genetic information is called genetic modification or genetic engineering. In the early 1980s, scientists developed recombinant DNA techniques that allowed them to extract DNA from one species and insert it into another. Refinements in these techniques have allowed identification of specific genes within DNA—and the transfer of that particular gene sequence of DNA into another species.

For example, the genes responsible for producing insulin in humans have been isolated and inserted into bacteria. The insulin that is then produced by these bacteria, which is identical to human insulin, is then isolated and given to people who have diabetes. Similarly, the genes that produce chymosin, an enzyme that is involved in cheese manufacturing, have also been inserted into bacteria. Now, instead of having to extract chymosin from the stomachs of cows, it is made by bacteria. This type of application of genetic engineering has not been very controversial. However, applications involving the use of plants have been more controversial.

Among the first commercial applications of genetically engineered foods was a tomato in which the gene that produces the enzyme responsible for softening was turned off. The tomato could then be allowed to ripen on the vine without getting too soft to be packed and shipped. As of 2002, over forty food crops had been modified using recombinant DNA technology, including pesticide-resistant soybeans, virus-resistant squash, frost-resistant strawberries, corn and potatoes containing a natural pesticide, and rice containing beta-carotene. Consumer negativity towards biotechnology is increasing, not only in the United States, but also in the United Kingdom, Japan, Germany, and France, despite increased consumer knowledge of biotechnology. The principle objections to biotechnology and foods produced using genetic modification are: concern about possible harm to human health (such as allergic responses to a "foreign gene"), possible negative impact to the environment, a general unease about the "unnatural" status of biotechnology, and religious concerns about modification.

BIOTECHNOLOGY IN ANIMALS

The most controversial applications of biotechnology involve the use of animals and the transfer of genes from animals to plants. The first animal-based application of biotechnology was the approval of the use of bacterially Scientists inserted daffodil genes and other genetic material into ordinary rice to make this *golden rice.* The result is a strain of rice that provides vitamin

A, a nutrient missing from the diets of many people who depend on rice as a food staple.

CONCERNS ABOUT FOOD PRODUCTION

Some concerns about the use of biotechnology for food production include possible allergic reactions to the transferred protein. For example, if a gene from Brazil nuts that produces an allergen were transferred to soybeans, an individual who is allergic to Brazil nuts might now also be allergic to soybeans. As a result, companies in the United States that develop genetically engineered foods must demonstrate to the U.S. Food and Drug Administration (FDA) that they did not transfer proteins that could result in food allergies.

When, in fact, a company attempted to transfer a gene from Brazil nuts to soybeans, the company's tests revealed that they had transferred a gene for an allergen, and work on the project was halted. In 2000 a brand of taco shells was discovered to contain a variety of genetically engineered corn that had been approved by the FDA for use in animal feed, but not for human consumption.

Although several anti-biotechnology groups used this situation as an example of potential allergenicity stemming from the use of biotechnology, in this case the protein produced by the genetically modified gene was not an allergen. This incident also demonstrated the difficulties in keeping track of a genetically modified food that looks identical to the unmodified food. Other concerns about the use of recombinant DNA technology include potential losses of biodiversity and negative impacts on other aspects of the environment.

Safety and Labelling

In the United States, the FDA has ruled that foods produced though biotechnology require the same approval process as all other food, and that there is no inherent health risk in the use of biotechnology to develop plant food products. Therefore, no label is required simply to identify foods as products of biotechnology. Manufacturers bear the burden of proof for the safety of the food. To assist them with this, the FDA developed a decision-tree approach that allows food processors to anticipate safety concerns and know when to consult the FDA for guidance. The decision tree focuses on toxicants that are characteristic of each species involved; the potential for transferring food allergens from one food source to another; the concentration and bioavailability of nutrients in the food; and the safety and nutritional value of newly introduced proteins.

VITAMINS

Vitamins are natural, even if manufactured, as long as they are labeled organic. They too have not been tested for safety at the mega doses that some faddists use them, though clinical studies have shown that they can be helpful when used intelligently. Obstetricians who frequently prescribe multiple vitamin pills for pregnant women and programmes for vitamin A intervention have saved children's eyesight and lives in poor countries.

The faddists who would regulate agricultural chemicals out of existence oppose even simple truth-in-labeling requirements for their beloved natural products. Groups such as the International Advocates for Health Freedom, the Life Extension Foundation, and "a growing number of allies in the 'patriotic' movement" see a secret conspiracy of "international pharmaceutical empires working hand-in-hand with the UN and the FDA to restrict access to lowcost herbs, vitamins and minerals around the world".

4

Food and Nutrition Biotechnology

THE RELATIONSHIP BETWEEN FOOD AND HEALTH

How healthy we are depends largely on what, how and how much we feed ourselves and what we take into our bodies consists of foods that sustain us and drugs that heal our dysfunctions and imbalances. Deep in our bodies, we are hosts to complex microflora, comprising a wide range of different bacterial species that play several roles: supplying their human host with additional value from foodstuffs; protecting against intestinal infections; and contributing to the development of the immune system.

Many health-improving properties of certain foodstuffs are already well known: dairy products may strengthen the immune system; fruits and vegetables contain vitamins that protect humans against infections; meat and fish deliver proteins important for the growth and development of the young body; fibre-rich foodstuffs are important for the intestinal transport of digested food; and several phytochemicals have a long-term protective function against cardiac diseases and, probably, cancer. Food safety as well as the health benefits from food pervading discussions in every sphere of society have become real, pressing concerns for consumers as they wonder whether the sources and objects of their dining pleasures are fraught with dangers to warrant their fear or constant vigilance.

OBESITY: A WORLD EPIDEMIC

In 2000, the World Health Organization produced a report that warned governments about a growing epidemic that threatened public health: obesity. In some countries, more than half the population is overweight, and in

December 2001 the US surgeon-general, David Satcher, gave a warning that obesity could soon kill as many people each year as cigarette-smoking. The World Health Organization general assembly, held in May 2004 in Geneva, had on its agenda a document entitled 'World Strategy for Food, Physical Exercise and Health'. Through this document, the WHO wanted to draw attention to the non-contagious diseases, which represent 60 per cent of world mortality and about 50 per cent of world morbidity. In addition to information and awareness campaigns, the WHO recommended a more stringent regulation on advertisement and labelling of foodstuffs, because 'consumers have the right to obtain correct, standardized and understandable information on the contents of foodstuffs, so as to make enlightened choices'.

The WHO's forecasts predicted that cardio-vascular diseases would be the first cause of mortality in developing countries by 2010, a status that is already the case in the industrialized countries. Atherosclerosis – a disease associated with the consumption of foods containing too much fat and sugars, a sedentary lifestyle and smoking – together with type-2 diabetes and obesity are real world epidemics. The increase in the number of persons suffering from type-2 diabetes is a matter of high concern.

The figure of 150 million patients may double in 2005 especially with the rise of those in pre-diabetic stages, characterized by intolerance to glucose and abnormal glycaemia before breakfast, as well as in the frequency of the metabolic syndrome. The latter is probably three to four times more frequent than the established type-2 diabetes, and it is a combination of obesity, an abnormal content of lipids in the blood, and hypertension. This syndrome is caused by an excess of body fat, especially in the abdomen, a sedentary way of life and inappropriate eating habits. In addition, the release of great quantities of free fatty acids by the body fatty tissue results in insulin resistance; as the activity of the hormone is inhibited, glucose cannot penetrate into the muscles and consequently glycaemia rises.

There is also the release by the fatty tissue of adipocytokines, anti-inflammatory substances that reduce the secretion of another hormone, adiponectin, which normally protects against insulin resistance and inflammation. Being overweight increases the risk of suffering from several related illnesses and may contribute to an earlier death. Women who are overweight run a risk five times higher than average of developing type-2 diabetes while those who are severely obese have a risk of more than 50 times higher. Obesity is also implicated in cancer: a recent study in USA showed that 14 per cent of cancer deaths in men and 20 per cent in women could be attributed to it.

Being overweight is also one of the main causes of heart diseases, the world's major cause of death malaria and AIDS. This problem does not seem less acute in the developing world. Asians and black Africans are even more susceptible to obesity and its related diseases than are Caucasians. For instance, 3 per cent of Chinese and 5.5 per cent of Indians are diabetic, compared with 3 per cent of British people. There are more new cases of diabetes in China and India than there are in the rest of the world put together. This is despite the fact that China was already spending 1.6 per cent of its annual gross domestic product treating non-communicable diseases, mostly obesity-related. The finger of blame seems to point to eating habits and also at the quality of foodstuffs. The trend in food manufacturing has been to produce cheaper food, which in some ways could have adverse human health effects. For instance, hydrogenated vegetable oil – vegetable fat made solid by adding hydrogen atoms – is the nutritionists' current enemy.

Widely used as a cheap substitute for butter and cream, it is the main dietary source of trans-fatty acids, heavily implicated in heart diseases. Some companies are therefore removing them from their products for fear of lawsuits. Cheap food may also make people eat more, and food companies certainly think giving people more food for their money makes them buy more. That is why portions of manufactured food and soft drinks have been growing in size and volume. Companies are now increasingly under pressure to stop selling to people more food for less money, but it is hard to reverse that trend. Tasty foodstuffs are generally sugary, fatty and salty.

Taste is as much instinct as habit, and once people are used to sugary, fatty and salty foods, they find it hard to give them up. Producing healthier foodstuffs that are also attractive to consumers' tastes could help solve the problem, in addition to education on better nutrition, food consumption habits and regular exercise. Health food is not a turn-of-the-21st-century invention. In 1985, people gave up caffeine; in 1987, salt; in 1994, fat. Now it is carbohydrates. But contemporary health-food consciousness may have stronger foundations.

The need for healthier food may also be a matter of demographics across timelines related to "demographic evolution" as the president of food system design at Cargill, Inc., pointed out. In 1975, there were 230 million over 65 years of age; 420 million in 2000 and 830 million was the estimate for 2025. As people become older, their willingness to spend money on staying healthy increases. Science has also contributed to the growing health-food consciousness. New Nutrition Business, a US consultancy firm, in 1996 there were 120 documents on nutrition science in peer-reviewed journals; in 2002, there were over 1000. With more scientific data, regulators are more willing to evaluate products and if so found with basis, allow health claims on products; and health claims increase sales.

The Atkins diet, during its peak days, which has boosted sales of eggs and meat, and hit potatoes, is one manifestation of consumers' determination to try various ways of programming their eating habits. Supermarkets also cater to this market. For instance, Waitrose's Perfectly Balanced Meals claim no more than 4 per cent fat, very little salt and no 'butylated hydroxanisole or hydroxytoluene' at all; and sales are rising at 20-25 per cent annually. Sales of nutritional supplements have more than doubled in the USA in the six years after the Food and Drug Administration liberalized labelling laws. In 2000, sales amounted to $17 billion and were increasing at 10 per cent a year. In the United Kingdom, by the end of February 2004, a report on public health commissioned by the government cited obesity among its main worries. Previous to that, the Prime Minister's strategy unit floated the idea of a 'fat tax' on foods that induce obesity; and in 2003, the Food Standards Agency – the industry regulation – advocated a ban on advertising junk food to children. Yet the UK government dismissed the idea of a fat tax, and the culture secretary stated she was skeptical about an advertising ban. The health secretary said the government wanted to be neither a 'nanny state' nor a 'Pontius Pilate state, which washes its hands of its citizens' health'.

Obesity Among Children

In France, obesity among children has been increasing since the early 1970s, particularly in the least-privileged social categories. The percentage of overweight schoolchildren has increased from 3 per cent in 1965 to 5 per cent in 1980, 12 per cent in 1996 and 16 per cent in 2003. The current figures are those prevailing in the USA during the 1970s, but the rate of increase is similar to that of the US. This illness has become a major challenge to public health and has been considered an epidemic by the French National Institute for Health and Medical Research. Jean-Philippe Ginardet of the Trousseau hospital in Paris, obesity among children is a frequent, serious and societal disease, difficult to treat, which leads, in the short term, to hypertension, diabetes and increase in the concentration of blood cholesterol.

It paves the way for cardio-vascular diseases among adults, *i.e.* for the first cause of mortality. Since 1992, evaluations have been carried out in schools of two cities in northern France. The first evaluation showed that children informed by their teachers had better nutritional knowledge and could therefore adopt better eating habits. The second evaluation, carried out in 1992 and 1997, revealed that within the families substantial change had occurred with respect to a better schedule of meals and to a significant reduction of animal fats in their diet. As a result, between 1997 and 2000, the incidence of obesity in the children in these cities has increased much less: +4 per cent among girls and +1 per cent among boys compared to the whole region that showed an increase of 95 per cent among girls and +195 per cent among boys.

This experimental approach to preventing obesity has lead to the launching of a five-year campaign named 'Together, let us prevent obesity among children' by the Observatory of Food Habits and Weight, and the Association for the Prevention and Treatment of Obesity in Pediatrics. Obesity is not a disease that is treated only with the assistance of physicians; it also concerns the family and society as a whole. While there may be basis to claim that the lack of exercise and the increasing time spent watching the television or using the computer, as well as junk food are considered important causal factors, obesity's etiology is not confined to lifestyles and habits. Family histories play an important role too, supported by the fact that 57 per cent of obese children have at least one overweight parent.

This underlines the genetic role as well as the conditions attending to the pre- and post-natal periods and to subsequent psychic and social factors in causing obesity. New epidemiological studies are needed to better understand the causes of the obesity epidemic. In France, a number of measures have been taken by the Ministry of Health within the framework of their National Programme for Nutrition Health, launched in 2001 and the nine priority objectives which aim at stopping the prevalence of obesity among children. These include: the distribution of food and education activities in some primary and secondary schools; setting up a working group on 'food advertisement and the child' with a view to reaching a compromise between the economic interests of the agri-food industry and public health constraints; recommendations to support breastfeeding; publication of a guide for children and teenagers on food and nutrition.

Physicians are requested to detect obesity as early as possible on the basis of reference graphs and a disk for measuring the index of body mass provided to them since November 2003. The WHO guide to measuring this index is as follows: the ratio of body weight to height rose to the power of 2; a resulting number above 25 is considered overweight and above 30 is "obese." These tools enable the physician to find out the period within which the accumulation of fat occurs – whether it is between the ages of 5-6 years and or before.

With only a 38% success rate of treatment among children, early detection of obesity may improve their chances. In Italy, since the early 1990s a centre has been working on the treatment of obesity among children in Atri, a small town of 11,000 inhabitants in the Abruzzes region. A recent survey in elementary schools showed that 31.6 per cent of children had a weight above the norm and 6.7 per cent of them were obese. Of the latter, the centre's physicians considered that only 5 per cent of obesity cases could be related to genetic or endocrine causes, while the rest were caused by bad eating habits.

It did not seem to be a question of quantity of food but of poor eating habits. Among these habits the physicians listed: the lack of breakfast, too many snacks composed of industrial foodstuffs, lack of, or very little consumption of fresh fruit and vegetables. The absence of exercise was also an aggravating factor. The treatment of obesity cases begins with the involvement of the family. Once a week, children should come to the centre with their parents and sometimes with their grandparents. In the centre's restaurant, a meal is served to them, containing pasta without fat, fish, fruit and vegetables. Children are not forced to eat meals to which they are not accustomed; they just have to try. The parents also eat the same meals. Then the children meet with the psychologist and nutritionist; the parents follow.

Family participation is crucial, because the parents should familiarize themselves with the carefully prepared and measured meals and above all they must understand that the children should not eat quickly, that pasta should not be left aside, that they should not eat while watching television, because this usually causes the child to lose control of what he/she eats.

The whole family should reconsider its way of preparing meals and eating them; that is why the centre's specialists insist that both children and grown- ups have their meals together and eat the same foodstuffs. During the summer, about 40 children between 7 and 10 years old are welcomed in a camp, located in a rural tourist centre seven kilometres from Atri. At the summer camp, children's nutrition is strictly controlled and physical exercise is a frequent practice, while television is prohibited. The objective is to consolidate the new relationship between children and their food. They learn how to identify foodstuffs through blind-tasting, *i.e.,* they develop their sense of smell and touch through handling them. It has been observed that children who attend the summer camp make remarkable progress with respect to their nutritional health and eating habits. This could be decisive in the treatment of obesity instant vigilance.

CHANGING EATING HABITS TO IMPROVE HEALTH AND WELL-BEING

People are consuming more and more food outside their homes. They eat in bars, restaurants, and other catering enterprises. The latest figures on the consumer barometre indicated that confidence in foodstuffs was undergoing a slow but sustained increase, in the European Union, with the notable exception of fast food. In the Mediterranean countries, the onslaught of fast food has destroyed good feeding habits but instead of the expected high obesity rate, the Mediterranean diet resulted in less cholesterol in the blood, and higher life expectancy.

But a study by Eurostat – the Statistics Centre of the European Union – warned that the South was no longer what it was. Not only have the Latins ceased to be slimmer than the Germans and the British. No less than 34.4 per cent of Greek men were overweight, as opposed to 29.5 per cent of their British counterparts and 28 per cent of Germans. The Greek population now possesses the highest proportion of overweight members among countries of the European Union, followed by Spain with 32%.

However, the Greeks had the lowest rate of dementia among the over 65's, and they still enjoy one of the highest life expectancies in the EU, with outstanding defences against colon cancer, hypertension and heart attacks. This maybe attributed to their high consumption rates for olive oil – 20 litres per person per annum – *i.e.* seven times more than the Spaniards'. Some years ago, attention was drawn to the 'Mediterranean paradox': Spain, France and Italy had fewer cardiovascular illnesses than their neighbours in Northern Europe, even though there were no significant differences in body weight. The difference lay in the diet, which includes abundant fruit and vegetables, olive oil as the main source of fat, more fish, the reasonable consumption of wine with meals and of generous inclusions of garlic, onions and nuts. However, in time, the greater consumption of meat and lesser consumption of vegetables, more sauces rather than oil and vinegar dressings, whisky and other spirits instead of wine, soft drinks instead of water, and a sedentary lifestyle have led to more digestive problems, higher blood pressure and more kidney failures and respiratory illnesses.

The Spanish sociologist and journalist Vicente Verdú, 'health has declined proportionally with the rise in the economy, and gastronomic ignorance has spread in pace with the cultural revolution. In the United Kingdom, there were signs that the problem of obesity was not necessarily worsening. For instance, while it enjoys the title of being one of the world's biggest consumers of chocolate, over the four years to 2002, sales of chocolate fell every year: 2 per cent by volume and 7 per cent by value over the period. In February 2004, the new chief executive officer of Nestlé Rowntree described it as 'a business in crisis'; although the company denied later on that there was a crisis, admitting only that sales of Kit Kat, its widely-known brand, fell by 2 per cent in 2003. Cadbury Schweppes, the United Kingdom's biggest producer of fattening foodstuffs, stated that five years ago, chocolate made it up to 80 per cent of sales; that was now down to a half. Five years ago, 85 per cent of sold beverages were sweet; that is now down to 56 per cent.\

The rest was mostly juice. Sale of diet drinks – which made up a third of the sales of fizzy drinks – have been growing at 5 per cent a year, while sales of fattening foodstuffs had been stagnant. In British supermarkets, people are buying healthier food. Tesco's director of corporate affairs, its Healthy Living range grew by 12 per cent in 2003, twice the growth in overall sales. Sales of fruit and vegetables were growing faster than overall sales, too. That may be partly because fresh produce is becoming more varied, there are more of them available all year round and better supply encourages more demand. Five years ago, Tesco stocked six or seven varieties of tomato, while nowadays it stocks 15.

A study carried out by the University of Southampton on a big new supermarket in a poor area of Leeds concluded that after it opened, two-thirds of those with the worst diets now ate more fruit and vegetables. Cafés and restaurants report an increase in healthy eating too. Prêt-A-Manger, a sandwich chain, stated that sales of salads grew by 63 per cent in 2003, compared with 6 per cent overall sales growth. Even McDonald's, which introduced fruit salad by early 2003, had sold 10 million portions since. There are also good signs in the area of physical exercise. Gym membership figures suggest that British people at least intend to be less indolent. Mintel, a market-research company, there were 3.8 million members of private gyms in 2003, up from 2.2 million in 1998. The overall results of these favourable trends was that the average man became thinner in 2002 while women's BMI was static, at least just as to body-mass-index which have only began to be recorded in 2002.

One year of course does not make a trend, but a decrease in America's weight in 2003, also for the first time, supports the idea that something is changing in the obesity trends of the two of the most developed countries in the world. On the other hand, where the rich lead, the poor tend to follow – partly because the poor become richer over time, and partly because health messages tend to reach the better-educated first and the less-educated later. That happened with smoking, which the rich countries gave up years ago, and the poor are nowadays trying to abandon.

As for government intervention in reducing obesity rates, campaigners for the "fat tax" point out that that this kind of intervention could aid the efforts to reduce obesity rates as government intervention did for smoking. But that may not necessarily be the case with food because consumers now are constantly assailed by messages from companies telling them to lose weight. Also, peer pressure among teens on weight issues may have more impact on teenagers than ministerial action. However, some forms of government intervention have triumphed.

For example, on 8 April 2004, the French parliament examined a bill that aimed at prohibiting automatic machines vending confectionery and soda in schools, and also on setting new rules on the advertisement of foodstuffs during television shows targeted to youth. On 30 July 2004, the French Parliament voted in favour of prohibiting as of 1 September 2005 vending machines in schools. This vote was cheered by 250 pediatricians and nutritionists working in hospitals who earlier on wrote to the minister of health a letter titled 'For a consistent nutrition policy of public health in France'.

The French traditional morning snack has been questioned. In January 2004, the French Agency for Food Sanitary Safety has published an advice against it; the Agency stated that the concern about compensating food insufficiency among a small minority of children leads to an unbalance of the diet of all schoolchildren; the additional food intake causes an excess of calories which leads to an increase in the obesity rate among children.

Vitamin-A Deficiency

More than 250 million children less than five-years old are exposed to the risk of vitamin-A deficiency worldwide. About 500,000 of them go blind annually and 2 million die from this deficiency every year. To address this deficiency, several strategies can be adopted: medical supplementation, *i.e.* prescribing vitamin-A pills; the enrichment of food with vitamin A in the agro- industry or when preparing food at the communitary level; and inducing the diversification of food resources that are locally available. The latter strategy was adopted in a pilot project carried out in Burkina Faso, West Africa, in conjunction with promoting the consumption of non-refined red palm oil.

From 1999 to 2001, in collaboration with researchers from the University of Montreal Department of Nutrition and from the Health Research Institute at Ouagadougou, Burkina Faso, scientists of the French Development Research Institute Unit on Nutrition, Food and Societies, have tested the efficacy of red palm oil on the body's vitamin A as it tested these on mothers and children under five years of age in the centre-east of the country, where this oil is not usually consumed. This oil, well known for its high content of beta-carotene - a precursor of vitamin A - is produced and mostly consumed in the north-west of Burkina Faso. Palm oil has therefore been transported to, and sold on, the sites of the pilot project in order to evaluate its impact on vitamin-A deficiency under conditions where women bought the oil freely and voluntarily.

The women were previously informed about the beneficial effects of red palm oil through debates, lectures, theatre performances, etc. The impact of palm oil was evaluated among women and children, at the beginning and the end of the pilot project through testing the amount of retinol in the blood serum. Results showed that after two years, the quantity of vitamin A ingested by the mothers and children who consumed red palm oil increased markedly: increase from 41 per cent to 120 per cent of safety inputs among the mothers and from 36 per cent to 97 per cent among the children.

Simultaneously, the proportion of mothers and children having a retinol content in the serum lower than the recommended threshold at the beginning of the study, has decreased from 62 per cent to 30 per cent for the women and from 84.5 per cent to 67 per cent for the children. These results demonstrated that red palm oil was an efficient food supplement in real commercial conditions for combating vitamin-A deficiency. In addition, about half of the women involved in the study modified their eating habits within two years while voluntarily consuming this foodstuff that was new to them. The consumption of red palm oil could therefore be incorporated, like other food items rich in provitamin A into national programmes for controlling vitamin- A deficiency in Burkina Faso, where the afore-mentioned pilot project is being extended, and in other countries in the Sahelian zone.

Artificial Sweeteners: the Case of Sucralose

Sucralose is an artificial sweetener that is arguably the food industry's hottest new ingredient, turning up in everything from the recently launched 'mid-calorie' versions of Coke and Pepsi, to low-carbohydrate ice cream. Yet this sweetener was actually invented in the 1970s. Its success has been the reward for the decades of toil by Tate and Lyle, the British ingredient-maker that patented the substance in 1976 and is currently selling it as a sugar substitute to food manufacturers. Johnson and Johnson, the US health-care group, is selling sucralose for home use under the brand name Splenda. The innovative sweetener is actually chlorinated cane sugar. During the manufacturing process, three hydrogen-oxygen groups on a sucrose molecule are replaced by three tightly bound chlorine atoms.

The resulting molecule is about 600 times sweeter than sugar and passes through the body without being broken down. The chlorinated agent is sodium chloride, and the underlying chemistry has not put off consumers or food manufacturers. Tate and Lyle has calculated that the worldwide market for 'intense sweeteners' was worth $1 billion a year at manufacturers' selling prices. In the relatively short time it has been available, sucralose has picked up 13 per cent of this market, giving it second place behind aspartame's 55 per cent, just as to Tate and Lyle.

Its compatibility with low carbohydrate dieting, not to mention direct praise from the late Dr Atkins himself, has helped. In the US, Splenda is now the leading sugar substitute, having surged ahead of the likes of Equal and Sweet'N Low. IRI, the Chicago-based market research company, revealed that Splenda accounted for 43 per cent of the sugar substitutes bought through US stores - excluding Wal-Mart - to May 2004. Sucralose is also making significant in-roads into the food-ingredients market.

Both The Coca-Cola Co. and PepsiCo., Inc., were using it in their new 'mid-calorie' colas, Coca-Cola C2 and Pepsi Edge, which have been designed to contain half the calories of the regular offering without diluting the sweetness as much as current diet versions. Because it performs better at staying sweet at high temperatures than other artificial sweeteners, sucralose can be used in foodstuffs that previously relied on sugar, such as microwaveable popcorn.

Because of its better sweetening performance at high temperatures, McNeil Nutritionals, the Johnson and Johnson's unit responsible for Splenda, was persuaded to introduce a bigger pack size for Splenda to cater for demand from bakeries. This 5lb 'baker's bag' retailed at $6.99-$7.99. Although the original patent dated back to 1976, sucralose had to wait until the 1990s for the first wave of regulatory approvals to come through. In 1991, it was cleared by Canadian authorities. Australia gave it the go-ahead in 1993. Tate and Lyle applied for US approval in 1987.

After a long time preparing all the technical information required for the application, US clearance was granted in 1998. In the EU, sucralose had already been available in the United Kingdom, Ireland and the Netherlands but only gained approval for use in all European countries by early 2004 after the publication of an amendment to the EU sweeteners directive. The swelling demand for sucralose led to speculation that the sole manufacturing plant in McIntosh, Alabama, might not cope. The factory used to be jointly owned by Tate and Lyle and McNeil Nutritionals, but the British company took full ownership in 2004 as they redrew their sucralose partnership. In June 2004,

Tate and Lyle announced the plant would be expanded at a cost of $29 million, the work being completed in January 2006. As for its safety to consumers' health, sucralose has faced claims spread through the Internet, as aspartame had been in its time, that it was not safe, in spite of obtaining official clearance in many countries. However, the Centre for Science in the Public Interest, a US lobby group noted for its scepticism of the food industry, declared that there was no reason to suggest that sucralose caused any harm.

PRODUCTION OF HEALTHIER FOOD

FUNCTIONAL FOODSTUFFS

The concept of 'functional' foodstuffs was defined in Japan by the mid-1980s. Japan had developed diet products with therapeutic properties. In many cases, these were fermented dairy products containing microorganisms having a favourable effect on the digestive tract and its processes. A functional foodstuff should be able to modify one or more organic functions favourably, in addition to its nutritional effect. For these products to be labelled as "nutraceutics" or "nutraceuticals", their therapeutic role should be demonstrated. These kinds of studies are of particular importance in the case of therapeutic claims against cancer and vascular diseases.

A precursor of nutraceutics is cod liver oil, which has greatly contributed to the control of rickets, a consequence of vitamin-D deficiency. In Europe and the USA, large-scale studies involving tens of thousands of volunteers are being carried out to determine the preventive action of vitamins A- and E- enriched substances and selenium-containing compounds on some pathological conditions resulting from the deficiency of these vitamins and selenium.

In the 1990s, the concept of the potential benefit of functional foodstuffs has become widespread, and the research carried out has led to its first products: an 'anti-cholesterol' oil, derived from maize; a rice deprived of its most allergenic properties; and a grapevine synthesizing more resveratrol. By mid-2003, David Sinclair - a pathologist at Harvard University - and colleagues reported in the journal Nature on resveratrol, a compound that could lengthen the life of a yeast cell by 80 per cent. Resveratrol activates enzymes that prevent cancer, stave off cell-death and boost cellular repair systems.

This naturally occurring molecule builds up in undernourished animals and plants attacked by fungi. Wine does not contain much resveratrol and the compound degrades in both the glass and the body. A pill might work better, and a provisional patent has been filed. D. Sinclair seems optimistic about the effect of resveratrol on extending human life expectancy. During the summer of 1999, Japan published a list of food of specified health use including 149 commercial products with a certificate from the Ministry of Health and Well- being. In the USA and Europe, the consumers can buy these pharma-foods or nutraceutics. The world nutraceutical market value was estimated at $50 billion in 2004. On 17 November 1999, Novartis AG announced the launching in Switzerland and the United Kingdom of a first line of nutraceutical products. Even if only several dozens of nutraceuticals are currently known, the nutraceutical industry is steadily poised to grow.

INDUSTRIAL PRODUCTION OF HEALTHIER FOODSTUFFS

Food science and biotechnology can lead to substantial innovations in the production of healthier foodstuffs as well as increased profits by major food companies as in the period 2003-2004. Consider Nestlé. The group is selling beverages, mineral water, dairy products, ice-creams, precooked meals, chocolate, pet food and cosmetic products. In 2002, Nestlé's annual turnover amounted to 87.7 billion Swiss Francs, broken down as follows: beverages, including mineral water; dairy products; precooked meals; confectionery; pet food; cosmetic products.

However, in 2003, net profit decreased to 6.2 billion SF, 17.3 per cent less than in 2002, owing to the weak economic growth in Europe and monetary fluctuations. Nestlé's biggest market is Europe with sales of 28.5 billion SF, followed by the American markets. The Asia-Pacific region also became a priority for the group's development with the turnover in that region reaching 14.4 billion SF in 2003. Nestlé spends 1.35 per cent of sales on research and development – a lot for a food company – and was employing 250,000 persons in 2003-2004 worldwide.

It explores the frontiers of nutrition research to determine what people should and should not be eating, to develop products such as milk with added long-chain polyunsaturated fatty acids and non-dairy products fortified with calcium for the lactose-intolerant individuals. Yakult – a bland, sweet, yellowish drink – is also a good example of industry that made good in healthy drinks. It is produced by the Japanese company Shirota, founded in 1955. Minoru Shirota discovered Lactobacillus casei shirota in 1930.

The product was launched in Europe in 1994 and since then has spread across the world. It claims the beneficial effects of lactobacilli on the intestinal microflora. It represented a $2 billion global business, and encouraged competition from other companies. Cargill, Inc., whose core business is commodities, employed 200 food scientists in 2003, up from 20 in 2000. It has developed many products with new ingredients, including Bon Appétit, a raspberry tea with soybean isoflavones, which 'may help promote bone health and relieve some of the symptoms of menopause'. While Kraft Foods and Cadbury Schweppes claimed they were removing some of the trans-fats out of their foodstuffs, PepsiCo, Inc., stated it has taken all the trans-fats out of its Frito-Lay snacks. This move was to a large extent the cause of a 30 per cent boost in fourth-quarter earnings. The drinks and snacks maker's quarterly profit was also lifted by lower costs associated with its 2001 merger with Quaker Oats. Fourth-quarter earnings were $897 million, or 51 cents a share, compared with $689 million, or 39 cents a share, in the same quarter a year earlier. Revenue rose 9.4 per cent to $8.1 billion.

The company continues to expand its snacks line with healthier offerings, *e.g.* new crisps, using maize oil rather than oil containing trans-fats. Frito-Lay's North American sales grew 6 per cent to $2.7 billion in the fourth quarter with volume up a smaller 3 per cent. The unit controlled almost two-thirds of the US snacks market. PepsiCo, Inc., is the world's fourth-biggest agri-food group, behind Nestlé, Kraft Foods and Unilever. In 2003, its turnover reached $26.971 billion and its net profit was $4.781 billion. Present in 160 countries, it had 140,000 employees.

The modification of vegetable oils is one of the key areas of plant and crop biotechnology, the overall objective being to increase their content in unsaturated fatty acids and to decrease that of saturated ones through conventional breeding, induced mutations or genetic engineering. Extensive work has been carried out on oilseed rape, soybeans, peanut and sunflower with good results that led to the commercialization of several products. Palm oil, which contains an equal proportion of saturated and unsaturated fatty acids, in addition to beta-carotene, is also a current research target, particularly of researchers at the Palm Oil Research Institute of Malaysia.

In addition, replacing triglycerides with diglycerides in vegetable oils render them free of trans-hydrogenated fats and good cooking oils, *e.g.* 'econa oil' in Japan. Inulin and oligofructosans refer to a group of fructose-containing carbohydrate polymers which, in many plant species, act as protective agents against dehydration and cold temperatures and also offer many health benefits to humans, mainly in the stimulation of the growth of beneficial micro- organisms called bifidobacteria.

These bacteria are sometimes used as a probiotic additive to foodstuffs such as yoghurt, as they can defeat harmful bacteria in the intestines and produce compounds with good health benefits. These dietary fructans are also reported to have a lipid-lowering potential. They are not digested in the upper gastro-intestinal tract and therefore have a reduced caloric value. They share the properties of dietary fibres without causing a rise in serum glucose or stimulating insulin secretion. Inulin and oligofructosans can be used to fortify foods with fibre or improve the texture of low-fat foods without resulting in adverse organoleptic effects.

Most of these two products currently on the market are either chemically synthesized or extracted from plant sources such as chicory roots. Oligofructosans are shorter chain polymers, highly soluble and provide 30 per cent to 50 per cent of the sweetness of sugar, and also have the other functional qualities of sugars. In formulation, inulin forms a smooth creamy texture, which makes this compound suitable as a fat substitute. We can also cite the work of F. Georges of the Plant Biotechnology Institute. He was working on the production of inulin and oligofructosans in separate transgenic plant experiments to compare the efficiency of their fibre production. Oilseed rape which is a poor producer of inulin and oligofructosans, was used as model system. In particular, the production of two enzymes was to be evaluated: sucrose-1-fructose-1-transferase which adds a fructose moiety to a sucrose molecule, and fructan: fructan fructosyl transferase which continues to elongate the polymer by adding more fructose moieties to the chain.

The study showed that both enzymes could be used in conjunction to produce inulins and oligofructosans. Growers of nutraceutical plants need varieties with good agronomic potential and those that are consistent with the varieties in terms of germination time, height and maturity. Growers will need to be able to guarantee the quality of their natural health-beneficial products. Breeding methods can therefore be used to achieve uniform quality for clinical testing and for product development, as well as to remove these potentially harmful or otherwise undesirable compounds that are produced in the plants along with their therapeutic ones.

To meet these goals, Alison Ferrie of the Plant Biotechnology Institute was using the doubled haploid technology or "haploidy", which facilitates the development of true-breeding lines. Immature pollen grains, called microspores, were cultured to produce haploid lines, whose genetic stock was thereafter doubled. True-breeding plants were thus produced in one generation, and doubled haploid techniques reduced the time required to develop a new variety by about three to four years. At the NRC-PBI, doubled haploid technology has been developed for oilseed rape and wheat. It is being applied to a wide range of nutraceutical and herbal species. Over 80 species have been screened for embryogenic response; anise, fennel, dill, caraway, angelica and lovage have shown good potential. Haploidy could also be combined with mutagenesis to enhance the desirable components or decrease the undesirable characteristics.

Mutagenizing single cells had definite advantages over seed mutagenesis. The new market for healthier foodstuffs attracts both the agri-food giants and pharmaceutical groups, so that the competition is harsh among them and the frontiers are less marked between both kinds of corporations. The competitive advantage of the food industry in this race is that it has a good knowledge of consumers' behaviour, massive marketing strategies while knowing that nutraceutics should remain tasteful and palatable if these were to be patronized by consumers.

In France, a success story was that of Danone's Actimel, launched in 1995 in Belgium in the form of a small bottle corresponding to an individual dose and commercialized in 15 countries. More than 600 million bottles had been sold worldwide in 1999, including about 100 million in France, where 9 per cent of the households of all socio-professional categories bought Actimel – dubbed the 'morning health gesture'. Others include that of the case of Eridania-Béghin Say in France in 1999, relating to food additives having an impact on cardio-vascular diseases, colon cancer, osteoporosis, diabetes, etc. which sold commercialized powder sugar enriched with 'biofibres', which boosts intestinal microflora and helps the body to naturally resist illness.

Back in Nestlé, they are also carrying out the relevant research-and-development work with the support of its 600-scientist strong nutrition centre, located in Lausanne while in May 1999, in the USA, Australia, and in Switzerland, Unilever with an international nutrition research centre at Vlaardingen, Netherlands, commercialized a 'hypocholesterol' margarine, which could help prevent the accumulation of 'bad' cholesterol. It also aimed to target markets in Europe and Brazil. In the USA, most agri-food companies have developed soups, beverages and cereals, which can help digestion and prevent cardio-vascular diseases and hypertension.

The US Food and Drug Administration has opened the way to nutraceutics, having labels carrying a health recommendation. On 21 October 1999, the FDA granted to soybeans the clearance to carry the claim 'may reduce cardiovascular risks' on their labels. This request was made by E.I. Dupont de Nemours and Co., Inc., the world's first-biggest producer of soybean products. Soya sauce and soybean paste are major foodstuffs across Asia. Industrial soybeans undergo a solid-state fermentation process using compliant stainless steel tanks instead of in conventional bamboo trays. They are also inoculated with Aspergillus oryzae selected strains that have been developed in Thailand to produce koji in higher yields and of better quality.

This technique, developed by a fermentation consortium associating the National Centre for Genetic Engineering and Biotechnology and the Department of Chemical Engineering of Kasetsart University has been successfully applied by the company Chain Co. Ltd., Bangkok, and thereafter adopted by some soya-sauce manufacturers in Thailand. The same company has succeeded in selecting the appropriate strain of Lactobacillus to replace the addition of acetic acid in order to enhance the sour taste of soya sauce. The company produces the top quality commercial soya sauce in Thailand – the so-called First Formulation.

The Case of Long-Chain Polyunsaturated Fatty Acids

Long-chain polyunsaturated fatty acids are a research focus for nutritionists and food biotechnologists. Their beneficial effect on the functioning of the cardio-vascular system has been initially mentioned since the 1970s in the medical literature. In France, David Servan-Schreiber – a psychiatrist advocating a 'medicine of emotions' – Guérir le stress, l'anxiété et la dépression sans médicaments ni psychanalyse has stressed the role of these fatty acids as anti-depression substances. Incidentally, the author is also a shareholder of a company that sells pills containing these fatty acids these long-chain polyunsaturated fatty acids belong to two main categories: omega-3 and omega-6. Among omega-3 fatty acids, there are the alpha-linolenic acid with 18 carbon atoms, eicosapentaenoic acid with 20 carbon atoms and docosahexaenoic acid with 22 carbon atoms.

The human body cannot synthesize the ALA as well as the linoleic acid which is an omega-6 fatty acid. Omega-3 fatty acids are found in rapeseed and soybean oils, marine animals and human milk. Food-consumption surveys carried out in France have shown that the consumption of omega-3 fatty acids was insufficient and the ratio of omega 6 to omega 3 was not balanced. Although research is being carried out on the precise role of these fatty acids on human health, it is not easy for the public to have a clear view of established scientific facts and amid controversial statements. Let us look now at what maybe causing confusion among the public as regard the issue of omega-3 fatty acids.

It may have begun with the study that revealed lower morbidity and mortality due to cardio-vascular of Greenland's Inuits who consume a lot of fatty fish. In France, the French Agency for Food Sanitary Safety convened a meeting of experts on the effects of omega-3 fatty acids on the cardio-vascular system. They concluded that the supplementation of daily diet with these fatty acids could have a beneficial impact on the functioning of the cardio- vascular system, as a secondary prevention measure. Morbidity and mortality reduction was indeed significant among the persons who suffered form cardio- vascular or metabolic diseases.

However, omega-3 fatty acids did not act on cholesterol; they may act on triglycerides and cell membranes, as well as on blood clotting and heart excitability; they may also have, through prostaglandins a positive effect on hypertension. The experts convened by the French AFSSA also warned against the role of the consumption of excessive quantities of omega-3 fatty acids, as they would increase cell susceptibility to free radicals. They recommended a maximum daily intake of EPA and DHA of 2g per day.

Then there are also the claims on the prohibitive effects of omega-3 fatty acids on tumors. To this, the AFFSA experts concluded that all the studies carried out up to 2004 on food habits did not substantiate in humans any evidence indicating that an enrichment of the diet with precursors of omega-3 fatty acids would protect against cancer. However, research work carried out on rats has shown that a diet enriched with omega-3 fatty acids caused a 60 per cent decrease in size of mammary tumours, twelve days after radiotherapy, compared with a 31 per cent decrease in animals fed with a non-enriched diet.

Trials are expected to be carried out on humans. Given the insufficiency of evidence, the benefits of taking Omega-3 pills remain inconclusive. In view of this, the general advice is to consume fish at least twice a week. The same goes for rapeseed oil. This is sufficient to meet the daily needs of omega-3 fatty acids. It is also recommended to feed poultry with rapeseed meal rather than with sunflower meal, because the former is richer in omega-6 fatty acids. Thus, consuming this kind of poultry meat would provide enough omega-6 fatty acids.

BIOFORTIFICATION OF FOOD CROPS

Biofortification of food crops makes sense as part of an integrated food-systems approach to reducing malnutrition. It addresses the root causes of micronutrient deficiencies, targets the poorest people, and is scientifically feasible and cost-effective. It is a first step in enabling rural households to improve family nutrition and health in a sustainable way. HarvestPlus is a coalition of CGIAR Future Harvest Centres or Institutes, partner collaborating institutions and supportive donors.

The International Centre for Tropical Agriculture and IFPRI are coordinating the plant breeding, human nutrition, crop dissemination, policy analysis and impact activities to be carried out at international Future Harvest Centres, national agricultural research and extension institutions, and departments of plant science and human nutrition at universities in both developing and developed countries. An initiative of the Consultative Group on International Agricultural Research, HarvestPlus is a global alliance of research institutions and implementing agencies coming together to breed and disseminate crops with improved nutritive value, *e.g.* with a higher content of iron, zinc and vitamin A. The biofortification approach is backed by sound science.

Research on this funded by the Danish International Development Assistance and coordinated by the International Food Policy Research Institute led to the following conclusions:

- Substantial, useful genetic variation exists in key staple crops;
- Breeding programmes can readily manage nutritional quality traits, which for some crops have proven to be highly suitable and simple to screen for;
- Desired traits are sufficiently stable across a wide range of growing environments; and
- Traits for high nutrition content can be combined with superior agronomic traits and high yields.

Initial biofortification efforts will focus on six staple crops for which prebreeding studies have been completed: beans, cassava, sweet potatoes, rice, maize and wheat. The potential for nutrient enhancement will also be studied in ten additional crops that are important components in the diets of those with micronutrient deficiencies: bananas/plantains, barley, cowpeas, groundnuts, lentils, millet, pigeon-peas, potatoes, sorghum and yams. During the first four years of the project, the objectives are to: determine nutritionally optimal breeding objectives; screen CGIAR germplasm for high iron, zinc and beta-carotene amounts; initiate crosses of high-yielding adapted germplasm for selected crops; document cultural and food-processing practices, and determine their impact on micronutrient content and bioavailability; identify the genetic markers available to facilitate the transfer of traits through conventional and novel breeding strategies; carry out in-vitro and animal studies to determine the bioavailability of the enhanced micronutrients in promising lines; and initiate bio-efficacy studies to determine the effect on biofortified crops on the micronutrient status of humans.

During the following three years, the objectives are to: continue bio-efficacy studies; initiate farmer-participatory breeding; adapt high-yielding, conventionally-bred, micronutrient-dense lines to select regions; release new conventionally-bred biofortified varieties to farmers; identify gene systems with potential for increasing nutritional value beyond conventional breeding methods; produce transgenic lines at experimental level and screen for micronutrients, test for compliance with biosafety regulations; develop and implement a marketing strategy to promote the improved varieties; and begin production and distribution.

During the last three years of the project, production and distribution of the improved varieties will be scaled up; the nutritional effectiveness of the programme will be determined; and the factors affecting the adoption of biofortified crops, the health effects on individuals and the impact on household resources will be identified.

Rice

Rice is the dominant cereal crop in many developing countries and is the staple food for more than half of the world's population. In several Asian countries, rice provides 50 per cent to 80 per cent of the calorie intake of the poor. In South and South-East Asian countries, more than half of all women and children are anaemic; increasing rice nutritive value can therefore have significant positive health impact. Food-consumption studies suggested that doubling the iron content in rice could increase the iron intake of the poor by 50 per cent; germplasm screening indicated that a doubling of iron and zinc content in unmilled rice was feasible.

Milling losses vary widely by rice variety, with losses of iron being higher than losses of zinc, which suggests than more zinc is deposited in the inner parts of the rice endosperm. Under the HarvestPlus project, improved rice germplasm will be provided to national partners in Bangladesh, Indonesia, Vietnam, India and the Philippines. The improved features will be incorporated into well-adapted and agronomically-preferred germplasm in ongoing breeding programmes at the national and regional level. A plant-biotechnology approach is the current priority for enhancing provitamin-A content of the rice endosperm. The leading varieties will be field tested for agronomic performance and compositional stability in at least four countries.

Wheat

The International Maize and Wheat Improvement Centre is leading the HarvestPlus research endeavour on wheat biofortification in order to increase people's intake of iron and zinc. Given that spring wheat varieties developed by CIMMYT and its partners are used in 80 per cent of the global spring wheat area, the potential impact of iron-enhanced wheat could be dramatic. The initial target countries will be Pakistan and India, in the area around the Indo- Gangetic plains, a region with high population densities and high micronutrient malnutrition.

The highest contents of iron and zinc in wheat grains are found in landraces of wild relatives of wheat such as Triticum dicoccon and Aegilops tauschii. Because these wild relatives of wheat cannot be crossed directly with modern wheat, researchers facilitated the cross between a high-micronutrient wild relative, Aegilops tauschii, and a high-micronutrient primitive wheat, Triticum dicoccon, to develop a variety of hexaploid wheat that can be crossed directly with current modern varieties of wheat and have 40 per cent to 50 per cent higher contents of iron and zinc in the grain than modern wheat.

The first biofortified lines will be delivered to the target region by 2005, *i.e.* broadly-adapted, high-yielding, disease-resistant wheat lines. The first high- yielding lines with confirmed iron and zinc contents in the grain should be available for regional deployment by mid-2007. Researchers will be exploring the introduction of the ferritin gene in wheat and will establish the feasibility of increasing the concentration of iron and zinc in the grain using advanced biotechnology approaches in addition to conventional plant breeding. Molecular markers for the iron and zinc genes that control concentration in the grain were being identified in order to facilitate their transfer. Scientists will also carry out studies on bioavailability to determine the extent to which iron and zinc status in animal and human subjects is improved when biofortified varieties are consumed on a daily basis over several months.

Maize

Maize is the preferred staple food of more than 1.2 billion consumers in sub-Saharan Africa and Latin America. Over 50 million people in these regions were vitamin A-deficient in 2004. The International Maize and Wheat Improvement Centre and the International Institute of Tropical Agriculture are identifying micronutrient-rich maize varieties and will carry out adaptive breeding for local conditions in partnership with National Agricultural Research Systems in Africa and Latin America.

The project under HarvestPlus is initially focusing on maize varieties having increased contents of provitamin A because a useful range of genetic variation has already been identified for this trait. The first target countries are Brazil, Guatemala, Ethiopia, Ghana and Zambia. To support the breeding programme, research is being conducted in Brazil, the USA and Europe to develop simple, inexpensive and rapid screening protocols for provitamin A, so as to reduce the cost of assays from $70-100 to $5-10 per sample. Research in Brazil and the USA is also focused on finding genetic markers to facilitate marker-assisted selection for provitamin A concentration.

In collaboration with the University of Wageningen, a human efficacy trial was planned with provitamin A-rich maize in Nigeria for 2005 in order to study provitamin A retention or loss for different storage, processing and common cooking methods. To facilitate extension and dissemination of biofortified maize varieties, country teams will be formed in the target countries in order to conduct adaptive breeding research, farmer-participatory variety evaluations, nutritional advocacy and promotional activities.

Beans

Common beans are the world's most important food legume, far more so than chickpeas, faba beans, lentils and cowpeas. For more than 300 million people, an inexpensive bowl of beans is the main meal of their daily diet. The focus of HarvestPlus research is on increasing the concentration of iron and zinc in agronomically superior varieties. Over 2,000 accessions from the International Centre for Tropical Agriculture gene-bank and several hundred collections of African landraces have been screened for their nutrient contents.

While the average iron concentration in these varieties is about 55 mg per kg, researchers have found varieties the content of which exceeds 100 mg per kg. The eventual goals are to obtain favourable combinations for productivity and nutritional traits, double the iron concentration and increase zinc concentration by about 40 per cent. The first bred lines with 70 per cent higher iron will likely emerge in 2006, while lines with double concentration of iron are anticipated in 2008.

The proportion of iron and zinc that can be absorbed from legumes such as common beans is typically low, due to anti-nutrients, specifically phytates and polyphenols, which normally bind to the iron and zinc, making them unavailable to the organism. Research indicated that it might be possible to reduce polyphenol concentrations genetically, thereby improving iron bioavailability.

In contrast, vitamin C is an iron-absorption enhancer because it binds to iron and prevents it from becoming attached to the iron-absorption inhibitors. Beans are often consumed with vegetables, including bean leaves with the potential of bean leaves as a source of vitamin C still to be explored.

Cassava

Cassava, also known as manioc or tapioca, is a perennial crop native of tropical America that is also widely consumed in sub-Saharan Africa and parts of Asia. With its productivity on marginal soils, ability to withstand disease, drought and pests, flexible harvest dates, cassava is a remarkably adapted crop consumed by people in areas where drought, poverty and malnutrition are often prevalent. Cassava is typically white in colour and, depending on the amounts of cyanogenic compounds, can be sweet or bitter. The International Centre for Tropical Agriculture will coordinate HarvestPlus' overall activities on cassava biofortification and be primarily responsible for research in Asia, Latin America and the Caribbean.

The IITA will be responsible for cassava biofortification in Africa. In collaboration with the University of Campinas, São Paulo State, Brazil, the total content of provitamin A in cassava varieties will be determined spectrophotometrically. Provitamin-A retention studies will also be carried out on different preparation and cooking methods used in cassava-consuming countries. A method for storing cassava roots for several weeks or a few months is needed for programmes quantifying hundreds of samples per year. Initial data suggest that the anti-oxidant property of a few yellow pigments in cassava roots may delay physical deterioration of the roots.

The longer shelf life of yellow cassava roots may not only appeal to farmers and consumers, but may also increase the demand for biofortified varieties. Nutritionally improved germplasm coupled with superior agronomic performance can be developed as a medium-term approach with products reaching the farmers as soon as 2009. The aim is to identify and select, from the varieties having both high provitamin-A contents and good agronomic performance, those with the highest iron and/or zinc content.

Sweet Potato

Sweet potato is an important part of the diet in East and Central Africa where vitamin-A deficiency is widespread. At present, African predominant sweet potato cultivars are white or yellow-fleshed varieties that contain small amounts of provitamin A. In contrast, the orange-fleshed varieties are believed to be one of the least expensive, rich, year-round sources of provitamin A. Boiled orange-fleshed sweet potato, such as the Resisto variety developed in South Africa, contains between 1,170 and 1,620 Retinol Activity Equivalents per 100 g and is estimated to provide between 25 per cent and 35 per cent of the recommended daily allowance for a preschool child.

Experts at the International Potato Centre who developed a biofortified orange-fleshed sweet potato, estimated that when fully disseminated, this sweet potato could reduce vitamin-A deficiency in as many as 50 million children. To encourage a switch from non-orange to orange-fleshed varieties, the texture of the latter must be changed because they tend to have a high-moisture content and adults prefer varieties with a low water content, *i.e.* a high dry biomass. Plant breeding is ongoing to increase the dry biomass of the provitamin A-rich orange varieties, to improve organoleptic characteristics and at the same time improve their resistance to viruses and drought. About 40 varieties of sweet potato with high dry biomass and provitamin-A content have been introduced to sub-Saharan Africa. Of these, 10 to 15 were being tested widely in different agro-ecological areas in some countries.

Some original varieties, mainly local landraces, have been well accepted by farmers and were being distributed on a small scale. HarvestPlus' biofortification activities in sweet potato will be initially focused on Ethiopia, Ghana, Kenya, Mozambique, Rwanda, South Africa, Tanzania and Uganda. The variation in provitamin-A content of newly harvested roots can be as much as 45 per cent. Much of the provitamin A appears to be retained during storage, food preparation and cooking. In the South African Resisto variety, the provitamin-A activity of the boiled roots was between 70 per cent and 80 per cent of that of freshly harvested roots. Additional studies were to be carried out in 2004 to determine the provitamin-A losses during food processing and cooking based on the usual practices found in East and Central Africa. A human bioefficacy study using an organoleptically acceptable promising variety was planned for 2005, once the food processing studies were completed.

The $100-million ten-year HarvestPlus programme will be financed during the first four years mainly by the World Bank, the USAID and DANIDA. The Bill and Melinda Gates Foundation would contribute $25 million towards the total cost of the programme. In addition, the Canadian Agency for International Development will allocate funds for the Latin American part of the programme.

REGULATORY ISSUES AND COMMUNICATION POLICIES

Innovation in healthy foodstuffs is also fraught with costly failures. For instance, Procter and Gamble spent 30 days developing Olestra, a fat which the digestive system cannot absorb. But the product has been dogged by claims that it inhibits the absorption of vitamins and nutrients that may help prevent cancer; by a hostile lobby group, the Centre for Science in the Public Interest; and by regulatory problems.

In 1996, after eight years of tests, the US Food and Drug Administration allowed it to be used as an ingredient, but products made with it had to carry the warning that it might cause gastro-intestinal distress. In the summer of 2003, the FDA allowed the warning to be taken off advertisements. Moves towards healthier products and functional foodstuffs also fuel the professional lives of lawyers, regulators and stock-market analysts.

Many companies have recently appointed advisory boards composed of top nutritionists. The FDA itself is acquiescing to companies' proposals to include in their products' health benefits in their labels. The FDA has, in fact, liberalized the rules on making health claims, adopting a four-tier system enabling consumers to decide based on how solid the science is behind any particular product's health claim. Calcium ability to protect against osteoporosis, for instance, is reckoned very solid while omega-3 fatty acids to prevent heart diseases are considered good, but second level.

By early 2003, the FDA announced that from 2006 consumers must be informed of the amounts of trans-fats in foods notwithstanding the already wide publicity on the adverse effects of transfats. Health authorities in Europe are also striving to regulate this kind of research and to establish marketing standards. All the difficulties relate to the need to demonstrate the impact of these foodstuffs on disease prevention in humans. Does this mean we have to wait for sixty years of clinical studies in order to obtain such evidence? Under the programme FUFOSE, Functional Food Science in Europe, priority is given to the determination of 'markers' which will scientifically enable the recognition of long-term benefits of functional foodstuffs. Those nutraceutics which will show a beneficial impact will receive an authorization for marketing, rather similar to that given to medicines. Genetic engineering is useful for producing crops or food ingredients deprived of some undesirable elements or enriched with healthy substances, and therefore qualified as nutraceutics.

To be attractive to the consumers, these foodstuffs should not be too expensive. Ageing populations are a particular target for nutraceutics, which can play a key role in the nutrition of old people suffering from under- and malnutrition. Between 1998 and 2002, it was estimated that the annual turnover of modified milks increased by 10 per cent in Europe and 36 per cent in the USA among people of more than 65 years. Agri-food companies are also designing communication policies not just for consumers but also for physicians, pediatricians and nutritionists, like the pharmaceutical groups, in order to highlight the benefits of their products. These policies have to take into account the cultural differences with regard to food and nutrition among the countries. They should also state the preventive role of nutraceutics as well as their therapeutic effects.

For example, information available to consumers regarding the LC1 yoghurt, which states that it contains bacteria that foster a balanced intestinal flora, has, just as to Nestlé, a more scientific slant in Germany, where one can talk of micro-organisms, while this approach would not be culturally accepted in France. This hints of cultural considerations in information about health products because despite the worldwide movement of people and international tourism, in countries of Anglo-Saxon culture food is generally considered as functional, *i.e.* one eats because he has to, while in the countries of Latin culture food must also give pleasure and should be surrounded with conviviality. Henceforth the need for communication policies to take account of this kind of nuances and that should be adapted to their targets.

PROBIOTICS AND PREBIOTICS

Probiotics are microbial food ingredients that beneficially influence human health while prebiotics are non-digestible carbohydrates such as fructo- and galacto-oligosaccharides. Probiotics have been used historically in different cultures in the form of fermented dairy foods, vegetables and cereals. Health effects of probiotics have been reported in the oral cavity, stomach, small and large intestine and the vagina. Although they consist mainly of lactic acid bacteria, bifidobacteria and yeasts have also been successfully used. Prebiotics are organic food components that exert health-promoting effects by improving the characteristics of intestinal flora. Established effects of prebiotics are dietary fibre-like effects such as anti-constipation, faecal bulking and pH reduction. The potential effects of prebiotics are similar to those of probiotics, since a major mechanism for prebiotics lies in the support of probiotics.

The synergistic combination of both probiotics and prebiotics is called symbiotics. At the Institute for Genomic Research in Rockville, Maryland, Karen Nelson and her colleagues at Stanford University are working to sequence the DNA of every bacterium found in the human gut. Previous estimates had put the number of bacterial species in the gut at about 500. Preliminary results from the guts of five healthy individuals have so far revealed about 1,300 species per person. As intestinal diseases could lead to subtle, long-term changes in gut bacteria, and changes in bacteria populations seemed to precede other diseases such as colon cancer, the TIGR team planned to compare healthy guts with those of people with Crohn's disease – an inflammation of the gut with no well-understood cause. On the other hand, sequencing the DNA of gut bacteria may lead to our being able to manipulate our intestinal flora in more sophisticated ways.

The TIGR researchers also planned to study the microflora of other parts of the human body, such as the mouth, skin and genitals, that each harbours its own distinctive community microbes. Under the European Commission, the cluster Proeuhealth brought together 64 research partners from 16 countries working in the fields of food, gastro-intestinal-tract functionality and human health.

The cluster aimed to provide: a clearer understanding of the relationship between food, intestinal bacteria, and human health and disease; new molecular research tools for studying the composition and activity of the intestinal microbiota; new therapeutic and prophylactic treatments for intestinal infections, chronic intestinal diseases and for healthy ageing; a molecular understanding of immune modulation by probiotic bacteria and examination of probiotics as vaccine-delivery vehicles; process formulation technologies for enhanced probiotic stability and functionality; and commercial opportunities for the food and pharmaceutical industries. Probiotic lactobacilli are known to affect immunomodulation.

The increased understanding of the molecular factors affecting immunomodulation and immunogenicity will allow the selection of probiotic strains, with enhanced protective or therapeutic effects. European researchers have targeted two types of intestinal diseases: inflammations such as inflammatory bowel disease and infections such as those caused by rotaviruses and Helicobacter pylori. Two specially-selected probiotics will be tested in long-term human clinical trials for their alleviating effects on inflammatory diseases, such as Crohn's disease and ulcerative colitis – immune-mediated diseases that result in chronic relapsing inflammation of the gut.

Delivering the health benefits of probiotics and prebiotics to consumers depends essentially on their successful processing. Viability, stability and functionality of these ingredients must be maintained during processing, formulation and storage. The effects of processing probiotics are being explored by the European researchers and used to develop optimal process and formulation technologies. New processing techniques will be applied to the development of functionally enhanced prebiotics and symbiotic combinations.

NUTRI-GENO-PROTEO-METABOLO-MICS ERA OF NUTRITIONAL STUDIES

Before the biotech era, research on food and nutrition dealt with establishing the importance of carbohydrates, fats and proteins in our diet, and with identifying trace elements and vitamins that are essential for the enzymes mediating our metabolism, respectively. The biotech era we are now just entering, concerns the understanding of the effects on specific gene expression of certain compounds we eat. A good example is that of the regulation of a diverse group of genes whose proteins, generally enzymes, tend to either directly or indirectly scavenge strong oxidants or to decrease the probability of production of strong oxidants; these proteins are referred to as phase-2 proteins and, since most of them are enzymes, they are commonly referred to as phase-2 enzymes. Strong oxidants can damage DNA thereby resulting in mutations that may lead to cancer; they scavenge the endothelial- derived vascular relaxation factor nitric oxide thereby promoting hypertension; and they also activate kinase pathways that lead to inflammation.

The phase-2 terminology comes from the terminology describing enzymes that metabolize xenobiotics: phase-1 enzymes being the mono-oxygenases, mostly cytochrome P450s, that convert the generally hydrophobic xenobiotics to strong electrophiles, while the phase-2 enzymes form water-soluble adducts by the addition of glutathiyl, glucuronosyl or sulphate groups.

Since the phase-2 enzyme genes all have anti-oxidant response elements in their promoter regions, any gene with an ARE in the promoter region is referred to as phase-2 protein. Phase-2 enzymes include the classical ones such as NAD(P)H: quinone oxireductase 1, glutathione S-transferases A, M and P families, UDP-glucuronosyl transferases, as well as the more recently defined phase-2 proteins: ferritin H and L chains; cystine/glutamate antiporter, peroxiredoxin I, heme oxygenase 1; L-gamma-glutamyl-L-cysteine ligase, metallothioneins, etc. All these proteins directly or indirectly inhibit strong oxidant formation, *e.g.* ferritin through sequestering iron, or promote strong oxidant scavenging, *e.g.* NAD(P)H: quinone oxireductase.

Phase-2 protein genes are coordinately upregulated through activation of an ARE in their promoter regions. Phase-2 protein inducers can be found in our diet: kaempferol, a flavonoid present in high amounts in kale; a flavonoid fraction found in blueberries/cranberries; enterolactone, a metabolite of the principle lignan secoisolariciresinol diglucoside found in the flax seeds; ellagic acid found in strawberries and raspberries/blackberries; the flavolignan silibinin obtained from milk thistle fruit; sulforaphane, the isothiocyanate metabolite of the glucosinolate glucoraphanin.

There is much evidence that dietary intake of such phase-2 protein inducers can increase phase-2 gene expression in a number of tissues and that such induction can decrease the incidence of chemically induced tumours. At the Plant Biotechnology Institute, Juurlink and colleagues are working on the phase-2 protein-inducing isothiocyanate derivatives of certain glucosinolates, 4-methylsulfinylbutyl glucosinolate, commonly known as glucoraphanin.

The Canadian researchers have shown that dietary intake of glucosinolates that give rise to phase-2 protein-inducing isothiocyanates can improve hypertension in the spontaneously hypertensive stroke-prone rats; in addition, oxidative stress and inflammatory changes in various tissues in the ageing SHRsp was down-regulated. The NRC-PBI's researchers also examined the effects of administration of the flavonoid quercetin in a neurotrauma model. They found that quercetin administration after spinal cord injury promotes retention of function, correlated with decreased inflammation. Not only is quercetin a very selective kinase inhibitor, but it is also known to be a phase-2 protein inducer.

In collaboration with Shawn Ritchie and Dayan Goodenowe, Juurlink has begun examination of the effect of broccoli sprouts containing glucosinolates that are converted into phase-2 protein gene-inducing isothiocyanates on the metabolic profile and they have seen pronounced effects in liver and other organs. In summary, we are entering an era in nutrition where we are beginning to understand how phytochemicals influence metabolism and gene expression.

Since many phytochemicals can have multiple actions such as activating signal transduction pathways that directly or indirectly alter gene expression or influence protein function that result in adverse metabolic reactions, one must use multiple approaches to understand how phytochemicals either individually or in combination affect us. Henceforth, a combined metabolomic/proteomic/genomic approach is required.

MODIFICATION OF FOOD TASTES AND HEALTHIER FOOD PRODUCTION

A breakthrough in the food industry would be to offer healthier versions of popular foodstuffs without affecting the taste. If it succeeds to do so, grapefruit juice could be sweet without added sugar and potato chips flavourful with half the current content of salt. This kind of research could have applications in medicine manufacture.

In April 2003, Linguagen Corp., a biotechnology company in Cranbury, New Jersey, conducting taste research, was granted a patent for the first molecule that will block bitter tastes in food, beverages and pharmaceuticals. The compound, adenosine 5'-monophosphate or AMP, occurs naturally and, when added to certain foodstuffs, including coffee and canned or bottled citrus juice, the company states, it blocks some of the acidic tastes from being felt by the tongue.

The finding of a bitter suppressor attracts all food companies, *e.g.* Coca-Cola Co., Kraft Foods and Solae, a soya-foods firm owned by E.I. Dupont de Nemours and Co., Inc., and Bunge have each expressed interest in flavour and taste biotechnology. Kraft Foods and Solae are Linguagen clients while Coca-Cola Co. has signed a research deal with Senomyx, another biotechnology company. Some research has focused on finding compounds that would trick the receptors on the tongue by accentuating or blocking certain elements in the food, allowing people to taste a cup of coffee without adding cream or sugar, or the sensation of full fat in low-fat products. Processed foods such as canned soups, sauces and snacks like potato chips contain high amounts of salt to mask the bitter tastes that result from the very hot cooking process.

Soft drinks are sweetened to tone down the bitter taste of caffeine. Food and beverage companies are, on the other hand, very concerned, as a group, about health and nutrition because of all the reports on epidemic obesity, epidemic diabetes, cardio-vascular diseases and hypertension. Hence the search for compounds that keep food tasty, minus salt, sugar and fat. So far scientists at Linguagen Corp. have discovered about 20 compounds that blocked bitter tastes and have been granted patents to use four of the compounds as bitter blockers. Because humans have more than 30 separate bitter taste receptors, finding a universal bitter blocker is nearly impossible. Linguagen Corp. is also trying to discover and market a natural sweetener to replace artificial ones like aspartame or saccharine, which often leave a bitter after-taste.

The company planned to license bitter blockers to food, beverage and medicine manufacturers in the USA by early 2004. Senomyx, based in La Jolla, California, is also developing bitter blockers, as well as molecules that block unpleasant smells and others that increase the salty taste in low sodium snacks while decreasing the product salt content. The research was in the early stages by mid-2003. The Coca-Cola Co. - the world's first-biggest soft-drink company, commercializing 400 beverage brands in 200 countries, with an annual turnover of $21.044 billion and a net profit of $4.347 billion in 2003 - is one of the company's clients. PepsiCo, Inc., is also interested in taste biotechnology and in anything that can impact food or beverages on a large scale.

Since AMP is not bioengineered and regarded as safe, it will be accepted by people and not shunned by consumers like previous additives which were supposed to revolutionize low-fat foodstuffs but later performed far below expectations. Much of current taste research is the result of radical rethinking of the mechanisms of perception of tastes by humans that has taken place since 1993. Researchers have shown that the human brain had the ability to recognize a variety of flavours including bitter, sour, savory and sweet all over the tongue rather than in specific areas of the tongue, as it was thought before.

The tongue papillae contain the taste buds; when food mixes with saliva, molecules dissolve on the papillae and, through the taste buds, send a signal to the brain, which interprets the flavour of what is being eaten. When a bitter blocker hits the tongue, it prevents the bitter taste receptors from being activated. The brain is thus unable to recognize the bitter flavour, while the latter is still embedded in the food or beverage.

CORRELATION OF GENETIC MARKERS WITH BEVERAGE AND FOOD QUALITY

CORRELATION OF GENETIC MARKERS WITH MEAT QUALITY

In 2002, the Maryland (Savage)-based biotechnology company, MetaMorphix, acquired the livestock genotyping business of Celera Genomics, a company founded in 1998 to sequence the human genome; it then joined up with Cargill, Inc., to commercialize a genetic test that will help to reveal, prior to slaughter, a cow's propensity to produce desirable meat. That task is being carried out by analyzing thousands of so-called single-nucleotide polymorphisms in the bovine genome. A SNP is a place where the genomes of individual animals vary by a single nucleotide. SNPs are therefore convenient marker versions of particular genes, and different versions of genes result in differences between animals. MetaMorphix and Cargill, Inc, tried to find out which SNPs were associated with variations in meat quality, such as flesh colour, amount of marbling, wetness and tenderness, so that these could be identified before slaughtering an animal, and suitable animals will thus be reserved for breeding.

In 2002-2003, Cargill, Inc., studied 4,000 cattle, trying to correlate MetaMorphix's genetic markers with meat quality - and with other important traits, such as growth rate. Almost 100 useful SNPs have been identified from this study. As a result, a prototype testing kit was to be used by the firm as of August 2004. The first 'designer meat' produced this way was expected to be marketed in 2005.

GENETIC TAGGING OF AQUACULTURAL SPECIES

Species-specific DNA markers can be used to identify animal species such as commercial molluscs and crustaceans, which represent a high proportion of aquacultural species. For instance, in Thailand, at the National Centre for Genetic Engineering and Biotechnology Marine Biotechnology Unit, species- specific markers based on 16S ribosomal DNA polymorphism have been developed for penaeid shrimps, tropical abalone and oysters. The black tiger shrimp is the most commercially important cultured species in Thailand. Because of outbreaks of diseases, the white shrimp has been introduced into Thailand and cultured commercially. On the other hand, external characteristics of P. monodon and P. semisulcatus are similar, but the growth rate of the latter is approximately three times slower than that of the former. In addition, P. merguiensis larvae, which could not yet be successfully cultured, were sold as those of P.vannamei. Species-specific markers were therefore developed for identifying the afore-mentioned species and P. japonicus as well. These markers can be applied to ensure quality control by properly labeling traded shrimp larvae.

Three species of tropical abalone are found in Thailand's waters: Haliotis asinina, H. ovina and H. varia. However, H. asinina is the most productive one, as it provides the highest ratio between meat weight and total weight. H. asinina specific markers based on 16S rDNA polymorphism have been developed in order to prevent supplying the wrong abalone larvae for the industry as well as to foster quality control of abalone products from Thailand. Oyster farming has shown rapid growth in Thailand over the last few years. Taxonomic difficulties relating to Thai oysters have had a limiting effect on the culture efficiency and development of their closed life cycle. Molecular genetic markers have therefore been developed to identify the three commercially cultured oysters, Crassostrea belcheri, C. iredalei and Saccostrea cucullata.

DNA FINGERPRINTING OF GRAPEVINE VARIETIES

Since 1990, US and French researchers have been trying to establish the phylogenetic tree of the varieties of grapevine grown throughout the world, using the fingerprinting technique. The research consists of analyzing the structure of certain regions of the genome of some grapevine varieties and comparing it with that of other varieties, so as to establish possible phylogenetic relations. For this analysis, the DNA is extracted from young ground leaves, but also from fruits and branches. The results of this first research have been published in 1999 in the Science journal.

There are about 6,000 grapevine varieties cultivated worldwide. In order to establish the origin of a variety, specialists used to rely on phenotypic traits, such as the morphology of leaves, berries and grapes. In this way, varieties could be grouped in a few families. DNA fingerprinting enables the researchers to go further. It appears that the current grapevine varieties are the remote offspring of the grapevines grown during the Antiquity around the Mediterranean, or during the European Middle Ages. The current varieties are the result of lengthy breeding activities, identification and comparison work, and stabilization of the selected strains or lines. The team led by C. Meredith and J. Bowers of the Department of Viticulture and Oenology of the University of California, Davis, has confirmed that the cabernet sauvignon variety – which is dominant in the Medoc region of France and is at the origin of most red grapevine varieties in the New World – was in fact the offspring of the cabernet franc and sauvignon, two varieties deeply entrenched in the middle valley of the Loire river.

The cooperation between the Californian team and the French specialists of the National Higher School of Agronomy, Montpellier, associated with the Genetic Research and Breeding Unit of the National Agricultural Research Institute has led to undisputable results concerning the origins of various grapevine varieties: the chardonnay, aligote, gamay and melon of Burgundy are all cousins and almost siblings. Such a conclusion, derived from the analysis of DNA fingerprinting, was not a surprise for oenologists and tasters because the wines made from these four varieties share common structures and aromas. This kinship is particularly expressed in the ageing wines.

Similarly, aged wines of the chenin variety resemble the Hungarian tokay. But more surprising than the discovery of that kinship among the four grapevine varieties, was the identity of one of the progenitors of the initial couple that gave rise to these varieties. Indeed, several historical elements were in favour of the creation of lines through the cross pollination between the pinot noir and the white gouais; these crosses have given birth – as proven by the fingerprinting analysis – to the three white and red varieties grown for a long time in various French provinces, and for some decades, in many regions of the globe.

The surprising aspect of this discovery was that the white gouais is almost unknown, although the vine specialists in Montpellier continue to grow it and to make wine form it for their own pleasure. However, this variety has played a key role in the origin of French viticulturc, just as to R. Dion in his Histoire de la vigne et du vin en France des origines au XIXe siècle. A document dated from the 12th century mentioned this variety as a lower-grade one; in 1338, the white gouais was found in Metz under the name of goez; at that time, instructions were given to eliminate this variety from all the Metz territory and to privilege only the white and black fromental, considered as higher- grade varieties.

The gouais was found in Paris during the 14th century and, owing to the expansion of the workers' population, it progressively replaced the pinot noir of Burgundy, which was a good variety of Parisian vineyards. The extension of the gouais was due to the wish of the winemakers to produce a cheaper wine. However, the phenomenon was limited to Paris and its suburbs; in the vineyards located away from the capital, the gouais was rejected, more noble grapevine varieties were used and contributed to the reputation of French viticulture.

The white gouais was also formerly grown in the Jura and Franche-Comté. For the US and French researchers, this variety which has played a key role in the history of vine and wine, is the same as the heunisch variety of Central Europe, introduced in Gaul by a Roman emperor originating from

Dalmatia. In Montpellier, the French researchers participating in the joint study with the US scientists from the University of California, Davis, have tried to reproduce the breeding between the pinot noir and white gouais in order to seek confirmation of the genetic research. Other attempts were expected to widen even more the range of cultivated grapevine varieties, for both their fruits and wines derived from them. But this approach was hindered by a drastic regulation, which practically prohibits any venture of this kind, while non-French winemakers and vinegrowers could do it.

In Apulia, in the heel of the Italian boot, drawing on grapes grown by up to 1,600 small farmers in the area, a California wine consultant associated with another Italian wine consultant from Friuli are producing and marketing wines that have scored a great success worldwide, with 2004 projected sales 15 times as big as those in 1998, the winery's first year. The wines are called A- Mano – handmade – and by far the best known is a robust red made from a once-obscure grape named primitivo. DNA testing by Carole Meredith at the University of California, Davis, established that primitivo is a descendant of a grape called crljenak kastelanski, widely known in the 18th and 19th centuries on the Dalmatian coast of Croatia.

California's zinfandel, she showed, is genetically the same as primitivo, though how it crossed the ocean remains a subject of dispute. Apulian primitivo and zin are not twins, of course; climate, soil and vinification all help to shape a wine's look, aroma and flavour, along with the grape variety. But the two share several characteristics: both are fruit-rich, chewy, sometimes lush wines, a deep violet-red in colour, often too high in alcoholic content for comfort, but much more subtle if carefully handled. For years, primitivo was used to add unacknowledged heft to chianti, barbaresco and even red burgundy. Nowadays, primitivo can stand on its own feet. In addition to A-Mano primitivo, other high-quality primitivos are grouped in an organization called the Academia dei Racemi, not a true cooperative but an association in which each member makes his own wine and joins the others for marketing support and technical advice. Based in Manduria, between the old cities of Taranto and Lecce, the group includes value-for-money labels like Masseria Pepe, Pervini and Felline.

FOOD SAFETY

It is an established fact that, despite current misgivings about food safety and unhealthy foodstuffs, what we eat and drink is nowadays subjected to more safety and quality controls than ever, and the effectiveness of these tests is demonstrated by the choices we regularly make. In Spain, for instance, just as to the 2003 Consumer Barometre released by the Eroski Group Foundation, the public has a degree of confidence in its food of 7.29 points out of 10. Furthermore, a survey by the Federation of Food and Drink Industries showed that 81.3 per cent of Spaniards interviewed considered the foodstuffs they bought as safe.

However, just as to a recent survey carried out by the Spanish Society for Basic and Applied Nutrition, 32 per cent of Spaniards had unsuitable daily eating habits and 64 per cent needed to improve their diet. By late 2002, the Spanish Agency for Food Safety had been created and its main task was to coordinate the implantation of effective control systems, with alert mechanisms to detect possible failures in the food-safety chain and manage them without repercussions on public health. The AESA is also acting as a watchtower for emerging risks and is responsible in Spain for handling alerts originated elsewhere.

During the agency's first year of existence, 633 food bulletins have been issued, 126 of which were alerts involving such administrative decisions as the withdrawal of certain batches of food. The AESA's president stated that 'we have to make sure that the consumer's perception of risk corresponds to the risk there actually is. Fear has no bounds, but information combats it on every front'. There were over 27,907 food industries registered in Spain, and the number of authorized abattoirs reached 800 in 2004. The food and drink industry turned over more than •600 billion a year, and the agricultural and food sector was the third-biggest employer in the European Union in 2003. Protective controls have to match up to this.

ORGANIC OR BIOLOGICAL AGRICULTURE

DEFINITION AND TRENDS

The phrase "organic or biological agriculture" designates an agricultural mode of production that does not rely on the use of chemicals, *e.g.* fertilizers and chemical pesticides. It also excludes any genetically modified organism, and is labour intensive. It does not mean that the products are necessarily of a higher quality than those derived from conventional agriculture. How much acreage is dedicated to this version of farming? It depends on the definition, but just as to an extensive survey by the Germany-based Ecology and Agriculture Foundation, in 2003 Australia led with 7.6 million hectares, followed by Argentina and Italy.

When expressed as a percentage of total arable land. The highest proportions are found in Europe where organic agriculture methods are well defined and products registered. Thus Austria, the leading European country in biological agriculture and products, devotes 11.3 per cent of its total arable land to this kind of agriculture. In France, the proportion is only 1.4 per cent and 11,000 farmers are considered organic producers. It is Europe's first-biggest agricultural country occupying 13th place with regard to biological agriculture.

The methods of biological agriculture have been applied in France since the 1950s, but it was only at the end of the 1970s that small groups of farmers began to organize themselves, *e.g.* to create the National Federation of Biological Agriculture. In October 1979, the first adviser in biological agriculture had been officially hired by an agriculture chamber, that of the Yonne Department, but the first farmers practising this type of agriculture were not taken seriously by their neighbours, or the institutions. During the summer of 1996, the 'mad-cow' disease led to the harsh criticism of industrialized agriculture. The French minister of agriculture highlighted the importance of high quality food and ordered a report on biological agriculture. At the end of 1997, the French government approved a multi-annual plan in favour of this type of agriculture.

The demand for its products was strong: in 1999, it grew by about 30 per cent, especially owing to the demand by the supermarkets, which showed a keen interest in the products. Despite government assistance provided since 1997 that aimed at facilitating the reconversion of farms, national production could not meet the market demand. In 1999, 8,140 farms were practising biological agriculture over 316,000 hectares, *i.e.* a threefold larger area than five years earlier, and this kind of cultivation concerned only 1 per cent of the whole useful acreage. To meet the demand, imports are growing, and it is forecast that in 2005, 1 million hectares should be devoted to biological agriculture in order to meet the demand. Specific training is being provided and financial assistance is given to those farmers who decide to carry out a reconversion to biological agriculture In the United Kingdom, demand for organic food is growing at over 40 per cent a year. This demand amounted to nearly $1.6 billion in 1999, with a third of the population buying organic food.

In France in 2002, 65 per cent of French people had consumed a 'bio' product during the last 12 months, compared with 50 per cent in 2001 and 40 per cent in 2000. In 2003, however, the French market of organic foodstuffs and products, labelled as "bio", after growing at a 20 per cent annual rate, showed signs of slowing down. Although those responsible for producing and transforming organic products tended to underestimate this stagnation, the growth rate of the 'bio' products market was estimated at 6 per cent to 10 per cent in 2003. The institute TNS Media Intelligence which scrutinizes the evolution of consumers' behaviour, 57 per cent of French housekeepers had bought at last one 'bio' product in 2003 - a low figure compared with the levels of consumption in the United Kingdom, Italy and Germany. There are several reasons for this stagnation. In the case of milk, the price of the litre has fallen by 0.02 to 0.08 cents since early 2002. Meat profits were 20 per cent to 30 per cent less.

Producers of poultry, vegetables and fruits suffered from crises comparable to those affecting their colleagues in conventional agriculture. However, there did not seem to be an overproduction, because although consumption of 'bio' products was decreasing, it was not met by French production. In 2002, ironically, 21 million litres of milk produced in France were 'declassified' and sold through conventional outlets, while 'bio' products sold in France were imported, *e.g.* fruits, vegetables, cereals and exotic products. Most of stakeholders involved in organic farming and production, as well as analysts, agree that the crisis is rather a marketing issue; in other words, 'bio'-product producers did not have the full capacity to put these products in the conditions when the consumers want to find them.

The French group Biolait that collects one-fourth of 'bio' milk produced in France failed to impose its prices to the transformers who went away; henceforth the declassification of this milk into conventional one, and the company Lactalis whose supply broke down, had to import its 'bio' milk from Germany. Meanwhile, there are also problems of distribution and price, as well as certification and fraud issues.

DISTRIBUTION OF ORGANIC OR 'BIO' PRODUCTS

Although hyper- and supermarkets are the first selling places of 'bio' products, the French Carrefour supermarkets sell about 300 products, while the British Sainsbury plc offer a much wider range of 1,200 products. Specialized shops are also less than 1,000 in France. Consequently customers have some difficulties finding 'bio' products in their usual shopping places. However, partnerships have been established between these retailers and the farmers. Some farmers are also trying to diversify their outlets so as not to depend on the supermarkets only: thus, in France BioBourgogneViandes, created in 1994 near Dijon and comprising some forty cattle raisers, is selling meat to individual clients directly. BioBourgogneViandes has also purchased a first butcher's shop in the central market of Dijon, while the number of members grew to 70 by the end of 1999, with an annual turnover of •2.4 million.

The cattle raisers owned four shops in different villages and were selling, in addition to meat such biological products as wine, cheese and vegetables. While in 1997, all the meat produced by BioBourgogneViandes was delivered to Auchan supermarkets, in 1999 only 40 per cent was bought by the latter and 60 per cent by individual customers through the associated butcheries and specialty shops. BioBourgogne Viandes claimed it had created jobs in small villages and saved businesses.

PRICING

The price of 'bio' products remains the principal obstacle to their purchase. Their survey carried out in October 2003 by the review 60 millions de consommateurs in France, 'bio' products sold in hyper- and supermarkets cost 40 per cent to 60 per cent more than conventional products, and 70 per cent to 100 per cent more in open-air markets and specialized shops. It is true that organic farming needs more labour, particularly for growing vegetables and fruits. Also the small volumes of milk and meat make the harvest, bottling and transportation of these products very costly, because one has to find the industrial tools that meet the stringent standards of this form of production. Only 'bio' eggs, laid by hens raised in open backyards and fed with foodstuffs derived from organic farming, have met with great commercial success.

There is also the issue of economies of scale. The high costs of 'bio' products and the need to successfully compete with conventional products have led organic farmers in Europe to request assertive policies from the respective governments in favour of organic agriculture, starting with a marked increase in the acreage devoted to it. Germany, Italy, Denmark and Austria have designed public policies to support the growth of organic farming, while in France it has been suggested that the acreage devoted to organic farming should reach 3 per cent of total arable land, in order to become economically sound.

CERTIFICATION

Biological agriculture in the case of France, is submitted to drastic constraints as determined by their agriculture ministry.

These are: culture rotation is strongly recommended in order to maintain soil fertility; animals should not be kept in narrow facilities:

- Breeds should be adapted to their environment and fed with products from biological agriculture, *i.e.* animal flours and GMOs are prohibited;
- Besides vaccination, animals are treated with the help of 'soft' medicine. Chemical fertilizers, herbicides, insecticides, synthetic fungicides are also prohibited. They are replaced by organic fertilizers, guano and marine algae, plant wastes and rock phosphate.

- Weeding is manual, mechanical or thermical.
- Pests are controlled through the use of nets, repellents and the release of natural predators.

Only when a great threat to the crops exists, chemical pesticides are authorized. for those who practise both conventional and biological agriculture, the fields devoted to each type must be separated and accountability should be distinct; and two or three years are also needed for converting farmland to biological agriculture so as to eliminate chemical residues in the soils. Products sold during this period are considered conventional, but the French government has decided to provide financial assistance in order to compensate the gap in revenue. In other words, organic farmers defend an alternative agricultural, economic and social model. They prohibit the use of genetically modified organisms and demand, in case the ban on transgenic crops is actually lifted in Europe, a threshold of 0.1 per cent for an adventitious presence of GMOs in their crops, instead of 0.9 per cent as it has been decided for conventional crops.

The requirements for organic farming are applicable throughout Europe, with some national differences that may create distortions in competition. In France, the requirements for livestock husbandry are the most stringent, as all the feed should be produced on the farm itself. The AB logo, which applies to 'bio' products in France, can be applied to non-French organic products provided that they meet the national requirements. A common European logo also exists to label those 'bio' products. These logos do certify the mode of production of the product, but not its quality as does the Red Label in France.

Regarding exotic products, they can also be labelled with the same AB logo, but they raise problems of traceability and control. Many developing countries are devoting an increasing acreage of their arable land to organic farming, as they are attracted by premium prices on the international market. This is, for instance, the case of Chile, which has important outlets for its organic products in the European Union and Japan. Surinam and Papua New Guinea head the list of African, Caribbean and Pacific countries as 'bio' producers. Demand in the Western countries for organic fruits and vegetables is enticing producers throughout the ACP countries to be organized and establish their foothold in the market of opportunity.

In early October 2001, more than 170 traders, producers, researchers and support agencies converged on Port-of-Spain, Trinidad and Tobago, to do that at a conference on diversifying regional exports through developing organic agriculture. They came from Cameroon, Malaysia, 17 Caribbean nations and departments, 11 countries of Central and South America, and eight countries of North America and Europe. The conference launched a new study on World markets for organic fruit and vegetables by the FAO, the International Trade Centre of the United Nations Conference on Trade and Development and World Trade Organization. Debates at the conference led to concrete proposals for national standards, regional certification, information services and special measures for smallholders wanting to switch to organic farming.

CERTIFIED DENOMINATION OF ORIGIN

The certification of organic or 'bio' products is part of a wider trend that consists of drawing the consumers' attention to the origin and quality of their foodstuffs and beverages. This trend responds to the concern that quality is threatened by industrialization of food production and processing. Although reaching a high quality standard may require years of work and great financial endeavours, farmers are interested in following suit. In France, for instance, one-third of farmers, 6,700 companies and 6,000 distributors were, in 2001, engaged in official procedures aimed at certifying their products.

Certification in France deals with the quality and origin of the product. In addition to the AB label for 'bio' products, there is the AOC label which dates back to 1935; it indicates the provenance of a product whose characteristics are associated with the natural environment and local knowledge. The Red Label, created in 1960, only indicates the higher quality of taste of the product, but not its provenance or mode of production; a transformed product with Red Label does not necessarily derive from a raw material of the same quality. Finally, a certificate of conformity guarantees, since 1988, the fulfilment of a series of requirements set up by a group of producers, processors or distributors; this private logo which entails a fee from the petitioner guarantees a constant quality of the product.

Over the 1990s, there has been a proliferation of mentions and signs, official or private, that claim the quality, origin or tradition of a wide range of agri- food products. This has been interpreted as the counterweight to the increasing industrialization of agriculture: in a few decades, the proportion of processed agricultural products purchased by households rose to 80 per cent and even more. On the other hand, food crises caused by listeriosis, bovine spongiform encephalopathy, foot-and-mouth disease, the suspicion about the effect of heavy metals on health as well as about genetically-modified organisms have provoked an anxiety among consumers who demand the traceability of their foodstuffs.

Farmers who have been submitted to the successive reforms of the common agricultural policy, wish to redefine their identity and role; they consider therefore that being able to trace their products and guarantee their quality is an appropriate way to respond to consumers' expectations. With regard to supermarkets, they have created brands which refer to the positive image of well-known production areas and which are recognized by the demanding consumer as a guarantee of quality.

To be granted an AOC, farmers must adopt a collective approach; they should create a union to defend the product, work together with the enquiry teams of the National Institute of Denomination of Origin during three, four or even ten years, and make a lot of efforts and use their know-how. From 1997 to 2001, the number of farmers having adopted this approach increased by 14 per cent, while the total number of farmers decreased by 4 per cent. The approach is rewarding: for instance, the production of a kind of cheese called 'morbier', which was granted an AOC in 2000, doubled in two years and the number of its manufacturers rose from 25 to 40; the price of the AOC cheese is 30 per cent higher than that of ordinary cheese. It is also true that granting of an AOC label and the quality attached to it have an impact on the value of land: the annual price of land planted with olive trees in the region of Nyons increased 2 per cent more than that of non-AOC land. This enabled the farmers to resist the pressure exerted on land by tourism or urbanization. AOC productions cannot be transferred outside their site of production. The green lentil of Puy or the Roquefort cheese is considered collective property and thus, cannot be expatriated. This is a major difference with a brand an industrialist can keep while transferring its production to Asia or Africa to lower manufacturing costs.

The trend towards quality and labelling has its limitations. Firstly, there is a risk of confusion among the consumers. The latter may be tempted to choose an imitation of AOC, less costly. It is the role of the INAO to monitor the market and make sure that the reputation of a product originating from a specific area is not undermined. It seems that in France the volume of AOC products on the market is close to what this market can absorb. For instance, AOC wines represent 55 per cent of total production; it is not considered unreasonable to raise this proportion and convert all the vineyards to that quality level. The same is true for poultry, a large part of which is being sold with a red label and an indication of origin.

The number of farms selling some 600 AOC products with a protected geographic indication was 140,000 out of a total of 650,000 on the French territory. These farms include vineyards, vegetable growers, fruit and olive-tree growers and livestock husbandry. Their number could still rise – one farmer out of three could be involved in this kind of production in the medium term – but this approach could not be extended to the whole French agriculture. European certification includes a certified protected origin, the French AOC, as well as the protected geographic indication, which establishes a less strict geographical relationship than the AOP. In the United Kingdom, a leading company in the production of 'bio' products is Duchy Originals, founded by the Prince of Wales and established at the Home Farm of Tetbury, near the Prince's residence of Highgrove.

It sells chocolates, bread, honey, biscuits, cheese, ham, sausages and soft drinks. On 23 March 2004, a new product has been launched: a 'bio' shampoo. The latter, bearing the logo Houmont, contains rose and lemon essences and will be a strong competitor of Body Shop's comparable products. It is a joint venture between the Prince of Wales and a famous London-based hairdresser. Duchy Originals is a prosperous enterprise, with a •22-million turnover in 2003-2004 and profit reaching •343,000.

The 'Prince Charles' label is found in big department stores as well as worldwide, particularly in the USA and Commonwealth countries. Benefits are transferred to a charity foundation chaired by the Prince of Wales. Since its creation in 1990, about £2 million had been transferred to philanthropic associations by Duchy Originals which employed 100 people. Official certifications are facing the competition of private certifications, such as the indication of local or regional brands, of provenance, emphatic mentions, which do not necessarily represent a qualitative content or value and which are not submitted to an independent review. Monitoring and control of 'bio' products are carried out in France by some thirty certifying bodies, except the AOCs which are under the control of the National Institute of Denomination of Origin. All these bodies are authorized by the state and their action is followed by that of the General Directorate for Competition, Consumption and Fraud Repression.

The denomination of origin label is not peculiar to rich countries. Thus, China, with the assistance of France, has adopted a law on appellations in 2000 and has created about 30 AOC labels concerning yellow wine, teas and hams that are typical for some regions. Vietnam has followed suit in 2001; the first geographic indication regarding the 'nuoc mam of Phu Quoc' has attracted the interest of Unilever which invested $1 million for transferring its production in that island; since then, the price of this AOC nuoc mam has trebled. Sixteen countries of West Africa have requested the French INAO to identify two products in each of them that could be certified. Morocco wants to protect its argan oil, while Bolivia wishes to label its wine and quiñoa – a nutritious seed from Amaranthus quiñoa – and Brazil wants to tag its best wines. These efforts demonstrate that the globalization of nutrition and food should not necessarily lead to homogenization of products but to the promotion of trade relations that respect nutritional differences and cultures.

SEGREGATION

In the fall of 1999, two French industrial corporations, Glon-Sanders – the leader in animal feed and egg production in France – and Bourgoin – the European leader in poultry production – decided to set up a non-genetically- modified soybean chain, through which this legume was traced during the whole transformation process, from the seeds to the eggs and poultry sold to retailers. Eleven cooperatives, including 2,000 farmers and representing 12,000 hectares, responded positively to both industrialists in order to produce the so-called 'soja du pays'.

In 1999, the expected harvest was 50,000 tons of beans and in 2000 more than 100,000 tons. The soybean meal or extruded beans would be used to feed poultry, sold under the Duc label as well as laying hens. It involved two production zones were concerned: one in the south-west of France, including eight cooperatives and the company, Céréol, which processes soybeans into meal; the second in Burgundy, including three cooperatives and the company Extrusel, specialized in the production of extruded seeds. In order to mitigate the risk of contamination of locally-produced soybeans by imported US beans, a cooperative from Castelnaudary the Groupe Occitan, checked the French origin of seeds, isolated the production plots, stored the harvested beans separately until they were delivered.

Regarding Extrusel, a subsidiary of four grain cooperatives of Burgundy and Franche-Comté and of two livestock-feed producers of Saône-et-Loire, it produced 20,000 tons of extruded soybeans per annum. Extruded soybeans are a very digestible feed which supplies both proteins and fats and is incorporated into poultry and hog rations. In February 1999, Extrusel made the decision to only use soybeans of which the non-GM status could be guaranteed. During the spring of 1999, 8,000 hectares were sown with soybeans that were certified as non-genetically-modified by about one thousand farmers; this acreage represented the whole cultivated area between the cities of Belfort and Mâcon. A fully operational traceability system was set up from the farm to the client. Part of the beans was produced in the company's station and the rest was bought from outside suppliers with guaranteed origin, so as to be in conformity with Extrusel ISO 9002 standards.

Farmers who subscribed to this new productive venture and who harvested their first non-GM soybeans in the 1999 fall, sold their product at 1,100 Francs per ton, compared with 1,000 Francs per ton of 'ordinary' soybeans and 1,300 Francs per ton of beans used for human consumption. Bourgoin imposed additional constraints on the farmers that produced the 'soja du pays'.

They must not use genetically modified soybeans and they should trace the production of the beans at all stages of the process, they must grow them at a certain distance of pollution sites and should refrain from spreading sludge originating from wastewater treatment. Certified poultry represented almost 85 per cent of annual total turnover of Bourgoin-Duc, and through the company's decision not to use transgenic soybeans it wished to anticipate consumers' demand. Bourgoin-Duc had already prohibited the use of transgenic maize since 1996 in the feed used for poultry. In addition Bourgoin had tried, with food distributors such as Carrefour, to set up non-GM soybean production chains in the USA and Brazil. In addition to Glon-Sanders' poultry and eggs, labelled as 'biologically produced' and qualified as high-quality and rather expensive products, French consumers could buy another type of product labelled 'soja du pays'. But for this kind of poultry, fed with non- GMOs and offering a good safety, the consumer had to pay more.

FRAUD

Biological agriculture is not free from criticism because of fraud. By early 2000, in France, the agriculture ministry's Directorate-General for Competition, Consumption and Fraud Repression carried out an enquiry on false biological cereals. About a dozen important operators were involved in the following traffic: a dealer buys conventional cereals and establishes forged certificates that qualify them as derived from biological agriculture; the cereals are sold as 'bio' products either to feed producers or directly to livestock raisers. As the selling price of 'bio' cereals could be twofold of that of conventional cereals, the illegal profit could vary from 1 Franc to 50 centimes per kilo, and because of the volumes concerned the benefits could be very high. The French authorities discovered an international network involving in particular Italian capital. Conventional cereals were sometimes purchased in France and shipped – really or virtually – to Italy, Belgium or the Netherlands, from where they returned with the 'bio' label. Another traffic was initiated in Central Europe, particularly in Romania or Ukraine, and the cereals were transferred to France.

On 23 March 2000, the French inspectors spotted a society based in

Brittany, Eurograin, which they suspected of having marketed 50,000 tons of cereals of doubtful origin in 1998 and 1999. Earlier, on 3 March 2000, a public enquiry had been opened in the Vienne Department regarding 12,000 tons of cereals commercialized by Bio Alliance, a company based in Chasseneuil-du- Poitou.

Its manager was condemned for having unduly used a 'bio' label on bovine meat. Another enquiry concerned the shipment of the Celtic Ambassador, a boat inspected in 1997 in Bordeaux; the 4,500 tons of cereals found in the boat officially originated from Romania, and had been certified 'bio' in the Netherlands. The enquiry showed that the shipment had been made at Fos- sur-Mer, in the southeast of France, and that the cereals were conventional French ones. Similar trafficking may crop up.

It underlines the limitations of the certifying bodies, in charge of controlling the fulfilment of biological agriculture requisites through two annual visits without warning. There were three certifying organisms in France. The most important one, Ecocert, covered 80 per cent of the market; by mid- 1999, it was able to detect pesticide traces in animal feed produced by Central Soya, a neighbour and client of Eurograin. Ecocert then alerted the certifying organism of Eurograin, Afaq-Ascert, which had been controlling this company since 1998. Many officials have recommended the stricter enforcement of biological agriculture requisites, considered as too loose. Fraud and the subsequent mistrust could also explain the relative slump in the consumption of 'bio' products. Another explanation of this decrease is the competition among quality labels as well as the trend towards a more environment-friendly agriculture that may kidnap the image relating to organic farming.

'RATIONAL' AGRICULTURE

"Rational" agriculture's goal is to make a compromise between productivity and environment conservation. In France, the terms of reference of this mode of agricultural production contain 98 points relating to the improvement of agricultural practices and upkeep of farm economic profitability. A key issue of conventional agriculture is the better and more effective use of fertilizers by crops. Increasing the absorption and assimilation of macronutrients such as nitrogen fertilizers would therefore contribute to decreasing the percentage of these fertilizers not used by plants and therefore to drastically reduce soil and water pollution by nitrogen compounds.

In this regard the work by Shuichi Yanagisawa and colleagues of the Universities of Okayama and Tokyo, published in the Proceedings of the

National Academy of Sciences on 18 May 2004 could be very promising. The Japanese researchers were able to incorporate into the genome of Arabidopsis thaliana a gene from maize that improves nitrogen assimilation in this crop species. The gene transferred, Dof1, does not only control the synthesis of a protein, but also a series of genetic regulations involved in the constitution of the plant's 'skeleton'.

In Arabidopsis thaliana, indeed, Dof1 has modified the expression of several genes: amino-acid concentration has increased, that of glucose was lowered compared with control plants, but not that of sucrose. But the most striking modification concerned the growth of the transgenic plants in an environment with limited amounts of nitrogen: control plants showed symptoms of deficiency such as blemished leaves, while transgenic plants looked normal. As a follow-up to the experiment on Arabidopsis thaliana, Dof1 has been transferred to potato plants by the Japanese workers; as a result, the amount of amino acids in the genetically modified plants increased.

This result is promising, because potatoes are not very efficient in absorbing and assimilating nitrogen compounds; henceforth the need to add such fertilizers to this crop. Increasing biological nitrogen fixation is also an objective of 'rational' agriculture. Legumes are able to fix atmospheric nitrogen thanks to symbiotic bacteria living in their root nodules and they need much less nitrogen fertilizers than other crop species. An international consortium is carrying out the sequencing of the genome of the annual alfalfa species, Medicago truncatula, considered as a model legume.

This genomics work is to be finalized in 2007, but preliminary results already showed that the genomes of legumes were remarkably conserved, *i.e.* the gene sequence is quite similar among the different species. About 500 genes have been identified as related to the symbiotic relationship between the legume and its rhizobia. Legume geneticists also hope to unravel the role of the Nod factor, a molecule identified by the researchers of the French National Agricultural Research Institute in Toulouse; it has been shown that one-tenth of mg of this substance per hectare was sufficient to raise soybean yields by 11 per cent.

Although the transfer of biological nitrogen fixation to cereals is considered a remote possibility because it would involve the transfer of a few hundred genes, the immediate priority is to select the most effective legume species and varieties using genetic methods, so as to raise the percentage of peas, lupins and horsebeans grown in Europe as feed and consequently decrease the dependence of European countries on the imports of soybeans for animal feed from the USA, Brazil and Argentina. Another objective of the selection of more effective nitrogen-fixing legume species and varieties is to reduce water and soil pollution caused by nitrogen fertilizers, as well as the fossil energy needed to produce these fertilizers: to produce and spray one ton of fertilizers, two tons of oil are needed. On 8 January 2003, at the Forum of 'rational' agriculture, its president, Christiane Lambert, indicated that between 5,000 and 10,000 pioneers of 'rational' agriculture were willing to move to this type of agriculture. However, the main trade union of non-intensive farming, opposed to transgenic crops, the Confédération paysanne, highlighted that half of the requirements of 'rational' agriculture in fact corresponded to just the compliance with current regulations.

The consumers' association UFC- Que choisir? did support the terms of reference, while the National Federation of Biological Agriculture was concerned about a confusion between the products of 'rational' agriculture and 'bio' products. Others fear that food processors and distributors may request the farmers to adopt 'rational' agriculture without any financial compensation, and may consequently neglect the farmers that do not move towards this mode of production. On 4 March 2003, a National Commission for Rational Agriculture and Qualification of Farms has been set up in order to gradually enlist the farms devoted to this type of agriculture.

While trying to regulate the so-called 'rational' agriculture, the French government wants to foster 'organic' agriculture. On 2 February 2004, the minister of agriculture announced a plan aimed at doing so. About •50 million over a five-year period were to be allocated to organic farmers. In addition, communication activities requiring a •4.5-million investment over three years aimed at clarifying the AB label which did not seem to be well understood by the consumer. This label means an alternative system of production that does not use chemicals, relies on antibiotics on a limited scale for livestock, that implies an extensive type of agriculture and includes crop rotations. It markets products that are certified to contain 95 per cent of ingredients derived from processes excluding the use of synthetic chemicals.

But the 'bio' label does not mean a superior taste or any health benefit. About •10.8 million over three years aimed at strengthening the downstream part of 'organic' agriculture, which is handicapped by the low volumes of production and the dispersion of producers on the French territory. In addition, biological or organic agriculture will be promoted in agricultural education and research. The National Federation of Biological Agriculture welcomed the governmental measures, but organic farmers stressed that they were not receiving financial aid aimed at this kind of agriculture in the very short term, like other European producers.

THE CASE OF SLOW FOOD: ORGANIC FARMING, EATING HABITS, TASTE AND CULTURAL FEATURES

The problems of organic farming may nevertheless be discounted by consumers from Europe or other reach countries on the basis of arguments relating to eating habits, taste and hedonism. Thus, Carlo Petrini, president of the association Slow Food is setting up in Italy in the heart of a 300-hectare domain the first world's 'university of taste'. By the end of 2003, some 400 students from the five continents were studying all the aspects of food culture in Piemonte. The ultimate goal is to train specialists that throughout the world will preach the art of good food.

Supported by the revenues from a hotel and a wine bank, the university will be completely independent from the food industry. The •16 million needed for the project have been found, especially from private savings. C. Petrini has built a real counterpower over 14 years, starting from a village in Piemonte, but now spreading outside the Italian borders. Slow Food's symbol is the snail – slow and tasty – and its slogan is: 'eat less and eat better'; it has also a web site, publishes a review in four languages and owns an editorial house. About 75,000 persons from Europe and North America adhere to the association and all defend the gastronomic heritage which is, just as to them, threatened by the homogenization of tastes, multinationals and hypermarkets. In October 2002, 140,000 persons attended the third Congress of Taste, organized by Slow Food in Torino.

The Slow Food's prize for the defence of biodiversity was awarded to 13 farmers from Japan, Greece, Guatemala and Guinea. They all had the merit to safeguard a product, such as an old rice variety, Andean vegetables or black piglets. Slow Food is difficult to define: it is a non-governmental organization, a consumers' association and a gastronomic club, and the whole managed as an advanced enterprise.

The objective is not to destroy private property or transgenic crop experimental plots, or to denounce steadily, but to play on pleasure, seduction and marketing, so as to successfully compete with fast-food companies and food and beverage multinationals. After having been launched in December 1989 in Paris, Slow Food was officially constituted with 500 persons from 17 countries. A few months later, the association named 150 good-will people whose task was to set up 'conviviums', *i.e.* autonomous clubs, where information and experience are exchanged, products are compared and tasted. In Milan, Slow Food's editorial house publishes guides and a luxurious cultural review on recycled document, that is sent to all the association's members.

In addition to the initial goal of eating less and eating better, there is also the concern for a better environment and food safety, *i.e.* 'produce less and produce better'. Surfing on the wave of preoccupation caused by the 'mad- cow' disease, Slow Food has gained momentum through its 'ecogastronomic' approach. In 1999, Slow Food launched the 'Taste Ark': foodstuffs and products threatened with extinction, once identified by the 'conviviums', are supported by various means, *e.g.* promotion tools provided by the association to the farmers or producers, exhibition booths at the Salon of Torino, assistance through the media, equipment and funds levied among sponsors and local authorities. More than 150 products have thus been saved, such as San Marzano tomatoes, Ischia's rabbit, argan oil in Morocco, lama husbandry in the Andes, as well as the relevant microeconomies. In 2004, some 300 products were to be saved.

Slow Food can claim that it has been successful in achieving its goal of advocating the importance of good food and the emphasis on maintaining and even widening the diversity of food culture, which entails an environment-friendly agriculture that includes organic farming. With 40,000 members in Italy, 9,000 in the USA, 6,000 in Germany and 3,000 in Switzerland, Slow Food is taken seriously by lobbies and big food and beverage companies. In France, the movement has taken root in the southern part of the country, but does not grow rapidly, probably because it is difficult to find a meaningful slot between gastronomy leaders and chefs, and the anti-GMOs vociferous opponents.

5

Food Borne Disease Examinations and Organize

FOODBORNE ILLNESS

A foodborne illness or food poisoning is any illness resulting from the consumption of food contaminated with pathogenic bacteria, toxins, viruses, prions or parasites. Such contamination usually arises from improper handling, preparation or storage of food. Good hygiene practices before, during, and after food preparation can reduce the chances of contracting an illness. The action of monitoring food to ensure that it will not cause foodborne illness is known as food safety.

Transmission

Some common diseases are occasionally transmitted to food through the water vector. These include infections caused by *Shigella,* Hepatitis A, and the parasites *Giardia lamblia* and *Cryptosporidium parvum.* Contact between food and pests, especially flies, rodents and cockroaches, are other food contamination vectors. Foodborne illness can also be caused by the presence of pesticides or medicines in food, or by unintentionally consuming naturally toxic substances like poisonous mushrooms or reef fish.

Symptoms and Mortality

Symptoms typically begin several hours after ingestion and depending on the agent involved, can include one or more of the following: nausea, abdominal pain, vomiting, diarrhea, fever, headache or tiredness. In most cases the body is able to permanently recover after a short period of acute discomfort and illness. However, foodborne illness can result in permanent health problems or even death, especially in babies, pregnant women (and their fetuses), elderly people, sick people and others with weak immune systems. Similarly, people with liver disease are especially susceptible to infections from Vibrio vulnificus, which can be found in oysters.

Incubation period

The delay between consumption of a contaminated food and appearance of the first symptoms of illness is called the incubation period. This ranges from hours to days (and rarely months or even years), depending on the agent, and on how much was consumed. If symptoms occur within 1-6 hours after eating the food, it suggests that it is caused by a bacterial toxin rather than live bacteria.

During the incubation period, microbes pass through the stomach into the intestine, attach to the cells lining the intestinal walls, and begin to multiply there. Some types of microbes stay in the intestine, some produce a toxin that is absorbed into the bloodstream, and some can directly invade the deeper body tissues. The symptoms produced depend on the type of microbe.

Infectious Dose

The infectious dose is the amount of agent that must be consumed to give rise to symptoms of foodborne illness. The infective dose varies according to the agent and consumer's age and overall health. In the case of Salmonella, as few as 15-20 cells may suffice.

Pathogenic Agents

An early theory on the causes of food poisoning involved ptomaines, alkaloids found in decaying animal and vegetable matter. While some poisonous alkaloids are the cause of poisoning, the discovery of bacteria left the ptomaine theory obsolete.

Bacteria

Bacterial infection is the most common cause of food poisoning. In the United Kingdom during 2000 the individual bacteria involved were as follows: *Campylobacter jejuni* 77.3%, *Salmonella* 20.9%, *Escherichia coli O157:H7* 1.4%, and all others less than 0.1%. Symptoms for bacterial infections are delayed because the bacteria need time to multiply. They are usually not seen until 12-36 hours after eating contaminated food.

Common bacterial foodborne pathogens are:

- Aeromonas hydrophila, Aeromonas caviae, Aeromonas sobria
- Bacillus cereus
- Brucella spp.
- ampylobacter jejuni which causes Guillain-Barré syndrome

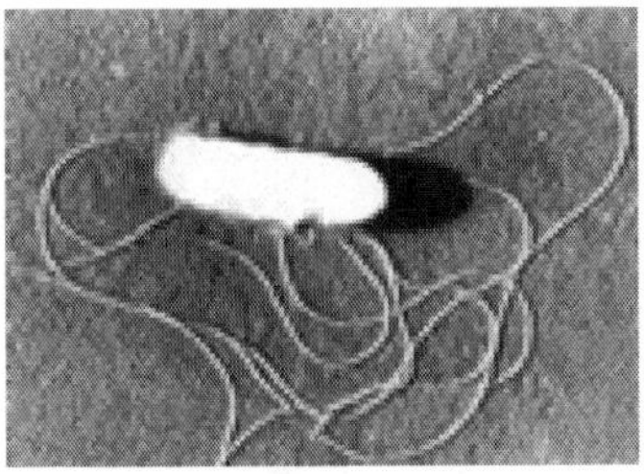

Fig. ***Listeria Monocytogenes***

- Corynebacterium ulcerans
- Coxiella burnetii or Q fever
- Escherichia coli O157:H7 enterohemorrhagic (EHEC) which causes hemolytic-uremic syndrome
- Escherichia coli - enteroinvasive (EIEC)
- Escherichia coli - enteropathogenic (EPEC)
- Escherichia coli - enterotoxigenic (ETEC)
- Escherichia coli - enteroaggregative (EAEC or EAgEC)

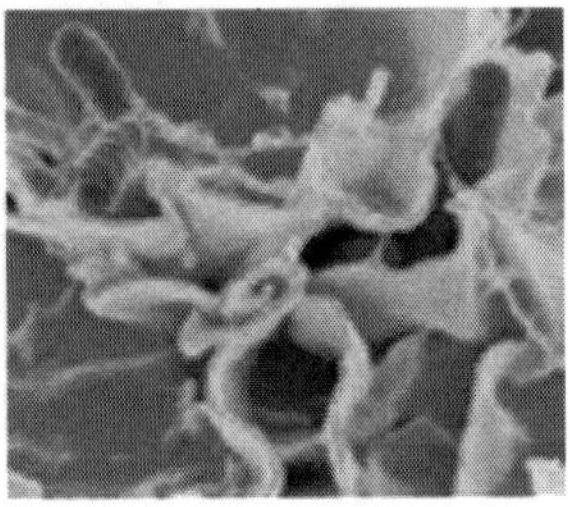

Salmonella

- Listeria monocytogenes
- Plesiomonas shigelloides
- Salmonella spp.
- Shigella spp.

- Streptococcus
- Vibrio cholerae, including O1 and non-O1
- Vibrio parahaemolyticus
- Vibrio vulnificus
- Yersinia enterocolitica and Yersinia pseudotuberculosis

Exotoxins

In addition to disease caused by direct bacterial infection, some foodborne illnesses are caused by exotoxins which are excreted by the cell as the bacterium grows. Exotoxins can produce illness even when the microbes that produced them have been killed.

Symptoms typically appear after 1-6 hours depending on the amount of toxin ingested:

- Clostridium botulinum
- Clostridium perfringens
- Staphylococcus aureus

For example *Staphylococcus aureus* produces a toxin that causes intense vomiting. The rare but potentially deadly disease botulism occurs when the anaerobic bacterium *Clostridium botulinum* grows in improperly canned low-acid foods and produces a powerful paralytic toxin.

Preventing bacterial food poisoning

The prevention is mainly the role of the state, through the definition of strict rules of hygiene and a public service of veterinary survey of the food chain, from farming to the transformation industry and the delivery (shops and restaurants).

This regulation includes:

- Traceability: in a final product, it must be possible to know the origin of the ingredients (originating farm, identification of the harvesting or of the animal) and where and when it was processed; the origin of the illness can thus be tracked and solved (and possibly penalized), and the final products can be removed from the sale if a problem is detected;
- Respect of hygiene procedures like HACCP and the "cold chain";
- Power of control and of law enforcement of the veterinarians.

At home, the prevention mainly consists of:

- Separating foods while preparing and storing to prevent cross contamination. (i.e. clean cutting boards, utensils, and hands after handling meat and before cutting vegetables, etc.)

- Washing hands and/or gloves before handling ready-to-eat foods.
- Respecting food storage methods (hot foods hot and cold foods cold) and food preservation methods (especially refrigeration), and checking the expiration date;
- Avoiding over-long storage of *left-overs*;
- Washing the hands before preparing the meal and before eating;
- Washing the fresh fruits and vegetables with clear water, especially when not cooked (*e.g.* fruits, salads), scrubbing firm fruits and vegetables with a brush to clean;
- Washing the dishes after use, rinsing them well in hot water and storing them clean and dry;
- Keeping work surfaces and chopping boards clean and dry;
- Keeping the kitchen and cooking utensils clean and dry;
- Not relying on disinfectants or disinfectant-impregnated cloths and surfaces as a substitute for good hygiene methodology (as above);
- Preventing pets walking on food-preparation surfaces.

Bacteria need warmth, moisture, food and time to grow. The presence, or absence, of oxygen, salt, sugar and acidity are also important factors for growth.

In the right conditions, one bacterium can multiply using binary fission to become four million in eight hours. Since bacteria can be neither smelled nor seen, the best way to ensure that food is safe is to follow principles of good food hygiene. This includes not allowing raw or partially cooked food to touch dishes, utensils, hands or work surfaces previously used to handle even properly cooked or ready to eat food.

High salt, high sugar or high acid levels keep bacteria from growing, which is why salted meats, jam, and pickled vegetables are traditional preserved foods. The most frequent causes of bacterial foodborne illnesses are cross-contamination and inadequate temperature control. Therefore control of these two matters is especially important. Thoroughly cooking food until it is piping hot, i.e. above 70 °C (158 °F) will quickly kill virtually all bacteria, parasites or viruses, except for *Clostridium botulinum* and *Clostridium perfringens*, which produces a heat-resistant spore that survives temperatures up to 100 °C (212 °F). Once cooked, hot foods should be kept at temperatures out of the danger zone. Temperatures above 63 °C (145 °F) stop microbial growth.

Cold foods should also be kept colder than the danger zone, below 5 °C (41 °F). However, *Listeria monocytogenes* and *Yersinia enterocolitica* can both grow at refrigerator temperatures. Hot foods should be held at 57°C (135

°F) or hotter until ready to cool. Hot foods need to be cooled quickly to limit the amount of time the food is in the danger zone (temperature range at which bacteria can grow.) The food should be cooled from 57 °C (135 °F) to 20 °C (70 °F) within two hours. Then further chilled to less than 5 °C (41 °F) in 4 hours. Food should then be held chilled at 5 °C (41 °F) or less.

Viruses

Viral infections make up perhaps one third of cases of food poisoning in developed countries. They are usually of intermediate (1-3 days) incubation period, cause illnesses which are self-limited in otherwise healthy individuals, and are similar to the bacterial forms described above.

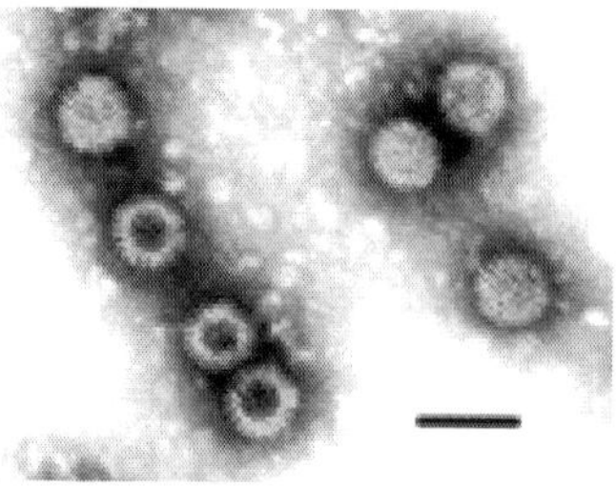

Rotavirus

- Norovirus (formerly Norwalk virus)
- Rotavirus
- Hepatitis A is distinguished from other viral causes by its prolonged (2-6 week) incubation period and its ability to spread beyond the stomach and intestines, into the liver. It often induces jaundice, or yellowing of the skin, and rarely leads to chronic liver dysfunction.
- Hepatitis E

Parasites

Most foodborne parasites are zoonoses. Platyhelminthes:

- Diphyllobo thrium sp.

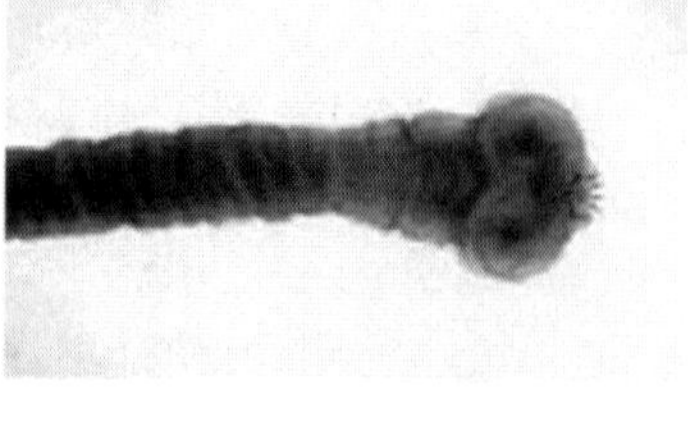

The scolex of *Tenia solium*

- Nanophyetus sp.
- Taenia saginata
- Taenia solium
- Fasciola hepatica Nematode:
- Anisakis sp.
- Ascaris lumbricoides
- Eustrongylides sp.
- Trichinella spiralis
- Trichuris trichiura Protozoa:

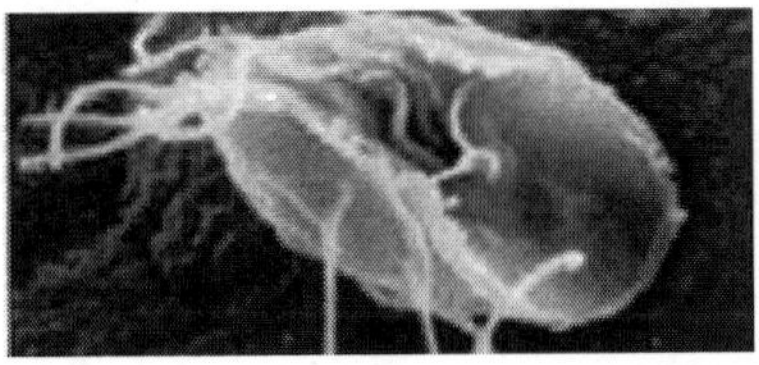

Giardia lamblia

- Acanthamoeba and other free-living amoebae
- Cryptosporidium parvum
- Cyclospora cayetanensis
- Entamoeba histolytica
- Giardia lamblia
- Sarcocystis hominis
- Sarcocystis suihominis
- Toxoplasma gondii

Natural Toxins

In contrast several foods can naturally contain toxins that are not produced by bacteria and occur naturally in foods, these include:

- Aflatoxin
- Alkaloid,

Fig. Red Kidney Beans

- Ciguatera poisoning
- Grayanotoxin (honey intoxication)
- Mushroom toxins
- Phytohaemagglutinin (Red kidney bean poisoning)
- Pyrrolizidine alkaloid
- Shellfish toxin, including Paralytic shellfish poisoning, Diarrhetic shellfish poisoning, Neurotoxic shellfish poisoning, Amnesic shellfish poisoning and Ciguatera fish poisoning
- Scombrotoxin
- Tetrodotoxin (Fugu fish poisoning)

Other Pathogenic Agents

- Prions, resulting in Creutzfeldt-Jakob disease

Statistics

There are every year about 76 million foodborne illnesses in the United States (26,000 cases for 100,000 inhabitants), 2 million in the United Kingdom (3,400 cases for 100,000 inhabitants) and 750,000 in France (1,210 cases for 100,000 inhabitants).

In the United States

In the United States, for 76 million foodborne illnesses (26,000 cases for 100,000 inhab.):

- 325,000 were hospitalized (111 per 100,000 inhab.);
- 5,000 people died (1.7 per 100,000 inhab.).

In France

In France, for 750,000 cases (1,210 per 100,000 inhab.):

- 70,000 people consulted in the emergency department of an hospital (113 per 100,000 inhab.);
- 113,000 people were hospitalized (24 per 100,000 inhab.);
- 400 people died (0.1 per 100,000 inhab.).

The causes of the illness (toxic factor) are:

Table. Causes of Foodborne Illness in France

Cause	Cases Per Year
1. Salmonella	~8,000 cases (13 per 100,000 inhab.)
2. Campylobacter	~3,000 cases (4.8 per 100,000 inhab.)
3. Parasites incl. Toxoplasma	~500 cases (0.8 per 100,000 inhab.)
	~400 cases (0.65 per 100,000 inhab.)
4 Listeria	~300 cases (0.5 per 100,000 inhab.)
5 Hepatitis A	~60 cases (0.1 per 100,000 inhab.)

The causes of death by foodborne illness are

Table. Causes of Death by Foodborne Illness in France

Cause	Cases per year
1. Salmonella	~300 cases (0.5 per 100,000 inhab.)
2. Listeria	~80 cases (0.13 per 100,000 inhab.)
3. Parasites	~37 cases (0.06 per 100,000 inhab.) (toxoplasma in 95% of the cases)
4 Campylobacter	~15 cases (0.02 per 100,000 inhab.)
5 Hepatitis A	~2 cases (0.003 per 100,000 inhab.)

Outbreaks

The vast majority of reported cases of foodborne illness occur as individual or *sporadic* cases. In most cases these originate, and occur, in the home. An outbreak occurs when two or more people suffer foodborne illness after consuming food from a contaminated batch.

Often, a combination of events contributes to an outbreak, for example, food might be left at room temperature for many hours, allowing bacteria to multiply which is compounded by inadequate cooking which results in a failure to kill the dangerously elevated bacterial levels.

Outbreaks are usually identified when those affected know each other. However, some are identified by public health staff from unexpected increases in laboratory results for certain strains of bacteria.

Ptomaine

"Ptomaine" is a former name for a supposed group of chemical substances that were theorized to cause food poisoning. The word "ptomaine" is no longer used scientifically.

Political issues

United Kingdom

Since the 1970s, key changes in UK food safety law have taken place following serious outbreaks of food poisoning. These included the death of 19 patients in the Stanley Royd Hospital outbreak; and the death of 17 people in the 1996 Wishaw outbreak of E.coli O157, which was a precursor to the establishment of the Food Standards Agency which, according to Tony Blair in the 1998 white paper *A Force for Change* Cm 3830 "would be powerful, open and dedicated to the interests of consumers". There remain questions, however, over the nature of any agency funded by a government which has not an insignificant say in staffing.

United States

In 2001, the Center for Science in the Public Interest (CSPI) petitioned the United States Department of Agriculture to require meat packers to remove spinal cords before processing cattle carcasses for human consumption, a measure designed to lessen the risk of infection by variant Creutzfeldt-Jakob disease. The petition was supported by the American Public Health Association, the Consumer Federation of America, the Government Accountability Project, the National Consumers League, and Safe Tables Our Priority.

This was opposed by the National Cattlemen's Beef Association, the National Renderers Association, the National Meat Association, the Pork Producers Council, sheep raisers, milk producers, the Turkey Federation, and eight other organizations from the animal-derived food industry. This was part of a larger controversy regarding the United States' violation of World Health Organization proscriptions to lessen the risk of infection by variant

Infections vs. Intoxications

- Infections characterized by
  - Viable cells consumed
  - Relatively long onset time for symptoms
  - Relatively long duration

  - Fever a common symptom (body's response)
- Intoxications characterized by
  - Viable cells may not be present
  - Quicker onset time (if preformed toxin)
  - Often short duration (unless toxin produced in body)
  - No fever (usually)
- BUT
  - Invaders make factors that help them adhere and penetrate
  - Toxins produced by many pathogens contribute to symptoms
  - Many organisms produce many different extracellular products (enzymes, toxins)
- SO
  - Intoxications are more common than you think!

Toxins

- Terms
  - Exotoxin = extracellular protein toxin
  - endotoxin = lipid A portion of Gram-neg outer memb.
  - enterotoxin = toxin that acts on gastrointestinal tract, producing typical food poisoning symptoms
- Nomenclature
  - Named for host cell attacked: cytotoxin, neurotoxin
  - Named for producer or disease: cholera, Shiga
  - Named for activity: lecithinase, adenylate cyclase
  - letter designation: exotoxin A
  - Types
  - A-B toxins
    a. Reduce temperature slow free radical
    b. B portion binds to receptor => specificity
    c. A portion enters cells, has enzymatic activity
    d. Examples:
    e. Cholera toxin
    f. e. coli LT (heat-labile toxin)
  - Membrane-disrupting toxins
    a. Nonenzymatic
    b. insert into membrane to form pores

c. Enzymatic attack on membrane components

d. Example: phospholipase (hemolysin)

– Superantigens

i. Cause hyperimmune response by the host Example: toxic shock toxin of S. aureus

Role of Toxins in Foodborne Disease

- Consumed as preformed ® self-limiting
- Produced by colonized bacteria ® local or distal effect
- Produced by infecting bacteria to aid invasion
- Autoimmune response to superantigens
- Endotoxin elicits immune response ® shock

How to Show an Organism Causes an Intoxication

- Feed culture and produce symptoms
- Feed culture filtrate and produce symptoms

Organisms Recognized for Foodborne Intoxications

- Staphylococcus aureus - enterotoxins
- Clostridium botulinum - neurotoxin
- Clostridium perfringens - enterotoxin, other toxins
- Bacillus cereus - diarrheagenic and emetic toxins
- Vibrio cholerae - cholera toxin
- enterotoxigenic E. coli - heat-stable and heat-labile toxins
- Aspergillus flavus, A. parasiticus - aflatoxins
- Fusarium sp. - fumonisin, zearalenone
- Aspergillus, Penicillium sp. - ochratoxins, patulin
- Alternaria - various toxins
- Gonyaulax catenella, G. tamarensis - saxitoxin

Agents of Foodborne Infections that also Produce Toxins

- Shigella sp. - shiga toxin
- Listeria monocytogenes - listeriolysin
- Salmonella sp. - enterotoxin, cytotoxin
- Enterohemorrhagic E. coli - shiga-like toxin
- Vibrio parahaemolyticus - hemolysin
- Yersinia enterocolitica - enterotoxin

- Campylobacter jejuni - enterotoxin
- Aeromonas hydrophila - multifunction toxin
- Plesiomonas shigelloides - enterotoxin

BACTERIAL AGENTS OF FOOD BORN ILLNESSES

STAPHYLOCOCCUS AUREUS

The Organism

- Gram-positive cocci
- Facultative
- Grows 10-45 °C (opt. 40-45)
- Grows pH 4.5 - 9.3 (opt. 6-7)
- Grows to aw 0.83
- Requires organic growth factors
- Salt-tolerant
- Heat-sensitive
- Doesn't compete well with normal flora of foods

Diseases

- Skin abscesses
- Scalded skin syndrome
- Toxic shock syndrome
- Nosocomial infections
- Food poisoning
  - Vomiting, diarrhea, nausea, cramps
  - Onset 1-6 hr
  - Recovery 1-3 days

Incidence in Environment and Foods

- Carried by animals and humans - skin, nasal cavity
- Foods of animal origin
- Foods receiving significant handling
- Frequently associated with outbreaks:
  - Baked ham
  - Pork
  - Salads

  - Pastries
  - Dried milk
  - Whey
- Used as indicator of proper handling, processing
- Much work done on detection and enumeration methods
  - Causes of sublethal injury
  - Methods to recover injured cells
- Risk factor: improper storage temperature
- Prevention: don't let organisms multiply in food

Extracellular Enzymes

- 34 different extracellular proteins
- Hydrolytic enzymes: proteases, lipases, hemolysins
- Penicillinase
- Coagulase
  - Clot formation in serum
  - Common identification test for *S. aureus*
  - Correlation with enterotoxin? < 100%
- Heat-stable nuclease
  - Attack RNA and DNA
  - Used to identify *S. aureus*
  - Correlation with enterotoxin? < 100%
- Toxins
  - Enterotoxins
  - Toxic shock toxin

Enterotoxins

- Merlin Bergdoll, Food Research Institute, Univ. Wisconsin
- Types: A, B, C1, C2, C3, D, E (SEA, SEB, etc.)
  - SEA most commonly found in outbreaks
  - SEB produced in high (but variable) amounts
- Structure
  - Single polypeptide chain
  - MW 26,360 - 28,500 (230 -239 amino acids)
- Mode of action
  - Stimulates vagus nerves in intestine?

  - Superantigen effects?
- Resistant to heat, radiation, proteolytic enzymes
- Toxic dose ~ 1 5g
- Production
  - Favored by optimum growth conditions
  - Produced during late log and stationary phases
  - need about 107 cells/g to get toxic dose

Toxin Detection

- Traditional animal assays
  - Monkey feeding
  - Injection of kittens i.p. or i.v.
  - Expensive, time-consuming, unpleasant
  - Sensitive, only measure of biological activity
- Immunological assays
  - Precipitation reactions (Ouchterlony, microslide)
  - Hemagglutination
  - ELISA, RIA
  - Problems with recovery from food and sensitivity
  - Must be able to recover and detect 1 ng/g food

Clostridium Botulinum

The Organism

- Gram-positive, anaerobic, spore-forming rods
- Defined and grouped by toxin production: A - G
- Grouped by physiology
  - Group I: proteolytic, high heat resistance
  - Group II: nonproteolytic, low heat resist., grow in cold
  - Group III: nonproteolytic
  - Group IV: little known

Occurrence in Environment and in Food

- Soil and sediment
- Fish
- Meats
- Fruits, vegetables, mushrooms

Growth Characteristics

- Temp. range: 10-50 C (aquatic strains grow to 3 C)
- Minimum aw: 0.94 (0.97 for aquatic strains)
- Minimum pH: 4.7-4.8
- Salt tolerance: up to 10% NaCl (6% for aquatic strains)
- Strict anaerobes
- Complex nutritional requirements
- Heat-resistant spores (especially type A)
- Don't compete well with normal flora
- Favored when heat has killed competitors

Potentially Dangerous Foods

- Heat treatment that kills vegetative cells but not spores
- Restricted oxygen (vacuum-packaged, modified atm)
- Low-acid
- Nutrient-rich

The Illness

- Categories
  - Foodborne
  - Infant botulism
  - Wound botulism
  - Unclassified
- Syndrome
  - Onset in 12-36 hr (avg)
  - Fatigue
  - Nausea and vomiting (early)
  - Poor coordination
  - Neurological signs
  - Respiratory failure
- Treatment
  - Try to clear toxin from GI tract
  - Antitoxin
  - Support therapy
- Prevention
  - Kill spores or prevent outgrowth

  - Specific processes for different food categories

The Toxin

- Structure
  - Progenitor vs. derivative forms (S, M, L, LL forms)
  - Monomer = 150,000 MW
  - Heavy chain: receptor and transfer functions
  - Light chain: endopeptidase activity
- Mode of action
  - Binds to and enters neurons at synapse
  - Interferes with release of neurotransmitters
- Genetic control
  - C1 and D carried by phage => phage conversion
  - G may be plasmid-borne
  - Others not located yet

Other Toxins

- C2 - may increase permeability of cells to toxin
- C3 - may affect some cell regulatory mechanisms

Lab Testing

- Look for presence of toxin
  - Directly in food
  - In cultures of organisms grown from food
- Mouse lethality test
  - Inject intraperitoneally
  - Trypsin treatment
  - Add antisera vs. A, B, E
  - Look for death in < 4 days
- ELISA, other immunological assays in development
  - Not as sensitive
  - Problems in toxin extraction
  - Don't detect activity

Clostridium Perfringens

The Organism

- Characteristics

  - Gram-positive
  - Sporeforming rod
  - Encapsulated
  - Anaerobic
- Types: A - E, based on toxin production
- Temperature effects
  - Growth range 15-50C (opt 43-45C)
  - High temperature:
    a. Vegetative cells inactivated >60C
    b. Low- vs. high-heat resistant spores
    c. Heat-activation for spore germination
  - Low temperature:
    a. Cold-sensitive vegetative cells
    b. Lose viability in stored samples
- O/R potential
  - Need anaerobic conditions for growth
  - Survive air exposure
- Aw: minimum ~0.95
- pH: optimum 6-7; range 5-8
- Salts: combined inhibition of NaCl + curing salts

Illnesses

- Foodborne disease
  - 8-24 hr onset
  - 24 hr duration
  - diarrhea + abdominal pain
- Pig-bel
  - Traditional pig feasts in Papua New Guinea
  - Type C (from pig) causative agent
  - 'Toxin (necrotizing) => bloody diarrhea
- Gas gangrene
  - Infection of deep wounds (anaerobic)
  - Tissue necrosis + gas production
- Antibiotic-associated diarrhea
  - Normal flora reduced
  - Resident *C. perfringens* proliferate

Foodborne Illness

- Reservoirs for the organism
  - Soil & dust
  - Intestine
  - Foods: meats, spices
- Outbreaks
  - Variable reporting
  - Frequently large (food service source)
  - Meat & poultry most common vehicles
  - Improper cooking, cooling
  - Prevention: temperature control
- History
  - Association with gas gangrene known since 1900
  - Food poisoning role suggested in 1940s
  - Study of organism in 1950s
  - Human feeding studies
    a. Remove off-flavors
    b. 1953: live cultures fed effective dose = 4-6 x 109 veg. cells
    c. 1971: ileal loop activity vs. human diarrhea
    d. 1977: 8-10 mg toxin induce symptoms neutralize stomach with bicarbonate

Pathogenicity

- Mode of action - macroscopic level
  - Fluid accumulation in intestine => diarrhea
  - Increased net secretion of fluid, Na+, Cl-
  - Glucose absorption inhibited
  - Damage to vili tips, brush borders
- Specific site of action
  - Disruption of plasma membrane
  - Toxin binds to membrane via 50,000 MW protein
  - Functional holes made
  - Altered membrane permeability
- Toxin
  - Protein, 34,000 MW, 309 amino acids
  - Activated by trypsin: 4000 MW peptide removed

- Released during sporulation
- Assays
  - Ileal loop
  - Animal diarrhea
  - Skin permeability
  - Serological

Bacillus Cereus
Theorganism

- Characteristics
  - Gram-positive
  - Sporeformer
  - Facultative
  - Very large cells
  - Related to:
- B. mycoides B. anthracis B. thuringiensis
  - Temp range 5-50C, opt. 35-40C
  - Short generation time (20-30 min)
  - pH range 4.5-9.3
  - Sporulates and germinates easily
  - High heat and drying resistance of spores
  - Acid from sugar fermentations (except mannitol)
  - Mildly proteolytic
- Occurrence
  - Wide distribution in soil, dust, air
  - Carried by humans and animals (from foods?)
  - In many food products:
    a. Dairy products
    b. Rice and cooked oriental foods
    c. Spices and spice mixes
    d. Dried products (flour, dry milk, pudding, soup mix)
    e. Beans and bean sprouts
    f. Meats
    g. Bakery products (cream-filled pastries)
- Extracellular products
  - Proteases (sweet curdling of dairy products)

- '-Lactamases (penicillin resistance)
- Phospholipases
- Hemolysins
- Toxins
  a. Lethal toxin: kills mice injected i.v.
  b. Bnterotoxins: diarrheal and emetic

Pathogenesis

- Infections in humans and animals
  - Wound infections
  - Septicemia
  - Mastitis
  - Meningitis
- Diarrheal food poisoning
  - Early reports in Europe in early 1900's
  - Linked to foodborne outbreaks in 1950's
  - Hauge (Norway)
    a. Lethal toxin: kills mice injected i.v.
    b. First complete account of outbreaks
    c. Drank culture in vanilla sauce -> symptoms
    d. Recognized worldwide by 1970's
  - U.S. outbreaks: 35 reported for 1977-1984
    a. Recognized worldwide by 1970's
    b. Meat loaf cooked rice
    c. Mashed potatoes green bean salad
    d. Chicken pot pie vanilla sauce
    e. Vegetable sprouts turkey loaf
  - High numbers needed (> 105/g)
  - Syndrome
    a. Vegetable sprouts turkey loaf
    b. Watery diarrhea, cramps, abdominal pain
    c. Nausea but not vomiting, no fever
    d. Onset 12-18 hr
    e. Duration <24 hr
    f. →similar to C. perfringens
  - Emetic food poisoning

a. First outbreak in 1971 (England, Chinese restaurant)
b. 110 outbreaks by 1979, most in Chinese restaurants or carry-outs
c. Implicated food: fried rice (boiled rice kept at r.t.)
d. Syndrome
e. Acute nausea and vomiting
f. Diarrhea not common
g. No fever
h. Onset 1-5 hr
i. Duration 6-24 hr
j. →similar to S. aureus
k. BUT B. cereus (high counts), not Staph, found in foods

Suveillance

- Toxins: no reliable assays available for routine use
- Organism
  - Ubiquitous → mere presence means little
  - High counts can indicate potential problems
  - How many isolates make toxins?
    a. Probably most make diarrheal toxin
    b. Unknown frequency for emetic toxin
- Methods
  - May need enrichment for routine screening
  - Selective/differential medium based on
    a. Resistance to polymyxin
    b. Inability to ferment mannitol
    c. lecithinase (phospholipase) production

Prevention

- Can't keep the organism out of foods
- Not dangerous at low levels => focus on control of growth
- Processing/cooking should
  - Destroy vegetative cells
  - Prevent spore germination and outgrowth
  - Prevent further multiplication
  - Thorough cooking should kill cells and spores

- Slow cooking, temps at 100C can allow spore survival
- Avoid nonrefrigerated storage of rice and other cereals
- Prepare food in small quantities for rapid heating, cooling

Diarrheagenic Types

- EPEC (enteropathogenic)
  - First pathogenic type identified
  - Severe illness in infants
- ETEC (enterotoxigenic)
  - Similarities to cholera
- EIEC (enteroinvasive)
  - Similarities to Shigella
- EHEC (enterohemorrhagic)
  - O157:H7
- EAggEC (enteroaggregative)
  - Newest identified type

ETEC

- History
  - Identified in 1950s: similar to cholera in ileal loops
  - 1960s and 1970s:
    a. Toxins identified (LT and ST)
    b. Epidemiology and prevalence travelers diarrhea, infant diarrhea diarrhea in domestic animals, esp. piglets
- Food and water transmission
  - Large inoculum (108) needed for illness
  - Usual vehicles: raw vegetables, ice, water
- Disease
  - Noninflammatory diarrhea, nausea, cramping
  - Mild (1-5 days) to severe (cholera-like)
  - Endemic in developing world - childhood diarrhea
  - Immunity in older children and adults - carriers?
  - Not common in industrialized countries
- Virulence factors
  - Colonization factors (CFAs)
    a. Needed to adhere to proximal small intestine

    b. Fimbriae mediate attachment
    c. Species specificity
  - Toxins
    a. Heat-labile toxin (LT)
    b. Heat-stable toxin (ST)

Cholera Toxin (A Digression)

- Mode of action
  - Colonization in the small intestine, toxin production
  - Normal intestinal function
    a. Fluid absorption/secretion controlled by ion conc.
    b. Specific pumps for Na+, Cl-, K+, HCO3-
    c. Net flow for ions + water is lumen -> tissues
  - Add organisms or cholera toxin
    a. No membrane damage
    b. Na+ absorption reduced,
    c. Cl- secretion increased
    d. Net water flow into lumen
- Structure
  - A-B type: 5 B-subunits + 1 A-subunit
  - Nick A to form A1 + A2 -> release A1 (active) part
- Host cell interaction
  - B-subunits bind to cell surface receptor (ganglioside)
  - A1 must enter cell and reach its target
- A1 activity
  - ADP-ribosylation of GS membrane protein
  - AMP production continues uncontrolled
  - Various transport processes affected by cAMP
- Genetics and control
  - Structural genes (ctxA, ctxB) in operon
  - May have multiple copies -> more toxin produced

Enterotoxins of ETEC

- LT (heat-labile toxin)
  - Inactivated at 60C for 30 min
  - 5 B + 1 A subunit C LT-I (humans, pigs) - very similar to cholera toxin

- LT-II (water buffalo) - different protein
- Same activity as cholera toxin
- Usually plasmid-borne
- ST (heat-stable toxin)
  - Active after boiling for 30 min
  - Small molecule (~2 kDa), internal cross-links
  - STa (human strains, methanol-soluble)
  - STb (porcine strains, methanol-insoluble)
  - Nonantigenic, not affected by anti-LT or anti-CT
  - Stimulates guanylate cyclase => accum. cGMP
- Detection
  - Animal assays ileal loop for LT suckling mouse for ST
  - Other assays for LT tissue culture - morphological changes immunoassays DNA probes

EHEC and Shiga-like Toxin

- Shiga toxin produced by S. dysenteriae
  - A-B type: 5 B subunits + 1 A subunit
  - A subunit affects ribosome -> protein synthesis stops
  - has enterotoxic, neurotoxic, cytotoxic activities, too

Salmonella Spp

1. *Name of the Organism*: Salmonella spp. Salmonella is a rod-shaped, motile bacterium — nonmotile exceptions S. gallinarum and S. pullorum—, nonsporeforming and Gram-negative. There is a widespread occurrence in animals, especially in poultry and swine. Environmental sources of the organism include water, soil, insects, factory surfaces, kitchen surfaces, animal feces, raw meats, raw poultry, and raw seafoods, to name only a few.

2. *Nature of Acute Disease*: *S. typhi* and the paratyphoid bacteria are normally caused septicemic and produce typhoid or typhoid-like fever in humans. Other forms of salmonellosis generally produce milder symptoms.

3. *Nature of Disease*: Acute symptoms — Nausea, vomiting, abdominal cramps, minal diarrhea, fever, and headache. Chronic consequences
   - arthritic symptoms may follow 3-4 weeks after onset of acute symptoms. Onset time — 6-48 hours. Infective dose — As few as 15-20 cells; depends upon age and health of host, and strain differences among the members of the genus. Duration of symptoms
   - Acute symptoms may last for 1 to 2 days or may be prolonged,

again depending on host factors, ingested dose, and strain characteristics. Cause of disease — Penetration and passage of Salmonella organisms from gut lumen into epithelium of small intestine where inflammation occurs; there is evidence that an enterotoxin may be produced, perhaps within the enterocyte.

4. *Diagnosis of Human Illness*: Serological identifica-tion of culture isolated from stool.

5. *Associated Foods*: Raw meats, poultry, eggs, milk and dairy products, fish, shrimp, frog legs, yeast, coconut, sauces and salad dressing, cake mixes, cream-filled desserts and toppings, dried gelatin, peanut butter, cocoa, and chocolate. Various Salmonella species have long been isolated from the outside of egg shells. The present situation with *S. enteritidis* is complicated by the presence of the organism inside the egg, in the yolk. This and other information strongly suggest vertical transmission, i.e., deposition of the organism in the yolk by an infected layer hen prior to shell deposition. Foods other than eggs have also caused outbreaks of *S. enteritidis* disease. It is estimated that from 2 to 4 million cases of salmonellosis occur in the U.S. annually.

The incidence of salmonellosis appears to be rising both in the U.S. and in other industrialized nations. *S. enteritidis* isolations from humans have shown a dramatic rise in the past decade, particularly in the northeast United States (6-fold or more), and the increase in human infections is spreading south and west, with sporadic outbreaks in other regions.

6. Relative Frequency of Disease Reported cases of Salmonellosis in the U.S. excluding typhoid fever for the years 1988 to 1995. The number of cases for each year varies between 40,000 and 50,000.

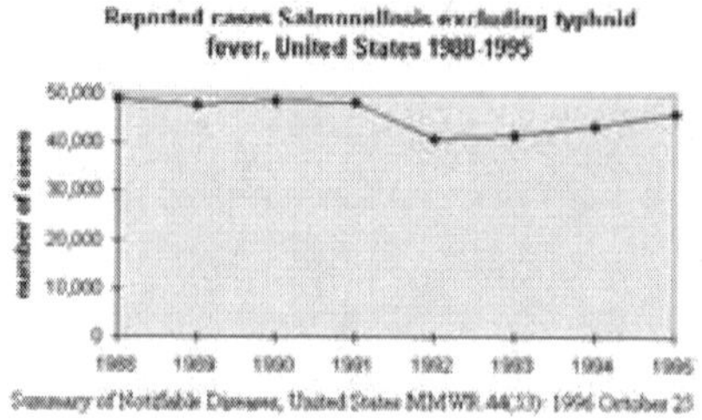

7. *Complications*: *S. typhi* and *S. paratyphi* A, B, and C produce typhoid and typhoid-like fever in humans. Various organs may be infected, leading to lesions. The fatality rate of typhoid fever is 10% compared to less than 1% for most forms of salmonellosis. *S. dublin* has a 15% mortality rate when septicemic in the elderly, and *S. enteritidis* is demonstrating approximately a

3.6% mortality rate in hospital/ nursing home outbreaks, with the elderly being particularly affected. Salmonella septicemia has been associated with subsequent infection of virtually every organ system. Postenteritis reactive arthritis and Reiter's syndrome have also been reported to occur generally after 3 weeks. Reactive arthritis may occur with a frequency of about 2% of culture-proven cases. Septic arthritis, subsequent or coincident with septicemia, also occurs and can be difficult to treat.

8. *Target Populations*: All age groups are susceptible, but symptoms are most severe in the elderly, infants, and the infirm. AIDS patients suffer salmonellosis frequently (estimated 20-fold more than general population) and suffer from recurrent episodes.

9. *Foods Analysis*: Methods have been developed for many foods having prior history of Salmonella contamination. Although conventional culture methods require 5 days for presumptive results, several rapid methods are available which require only 2 days.

10. *Selected Outbreaks*: In 1985, a salmonellosis outbreak involving 16,000 confirmed cases in 6 states was caused by low fat and whole milk from one Chicago dairy. This was the largest outbreak of foodborne salmonellosis in the U.S. FDA inspectors discovered that the pasteurization equipment had been modified to facilitate the running off of raw milk, resulting in the pasteurized milk being contaminated with raw milk under certain conditions. The dairy has subsequently disconnected the cross-linking line. Persons on antibiotic therapy were more apt to be affected in this outbreak. In August and September, 1985, *S. enteritidis* was isolated from employees and patrons of three restaurants of a chain in Maryland. The outbreak in one restaurant had at least 71 illnesses resulting in 17 hospitalizations.

Scrambled eggs from a breakfast bar were epidemiologically implicated in this outbreak and in possibly one other of the three restaurants. The plasmid profiles of isolates from patients all three restaurants matched. The Centers for Disease Control (CDC) has recorded more than 120 outbreaks of *S. enteritidis* to date, many occurring in restaurants, and some in nursing homes, hospitals and prisons. In 1984, 186 cases of salmonellosis (*S. enteritidis*) were reported on 29 flights to the United States on a single international airline. An estimated 2,747 passengers were affected overall. No specific food item was implicated, but food ordered from the first class menu was strongly associated with disease.

Campylobacter JejunI

1. *Name of the Organism*: *Campylobacter jejuni* (formerly known as

Campylobacter fetus subsp. jejuni) *Campylobacter jejuni* is a Gram- negative slender, curved, and motile rod. It is a microaerophilic organism, which means it has a requirement for reduced levels of oxygen. It is relatively fragile, and sensitive to environmental stresses (e.g., 21% oxygen, drying, heating, disinfectants, acidic conditions). Because of its microaerophilic characteristics the organism requires 3 to 5% oxygen and 2 to 10% carbon dioxide for optimal growth conditions. This bacterium is now recognized as an important enteric pathogen. Before 1972, when methods were developed for its isolation from feces, it was believed to be primarily an animal pathogen causing abortion and enteritis in sheep and cattle. Surveys have shown that *C. jejuni* is the leading cause of bacterial diarrheal illness in the United States. It causes more disease than *Shigella* spp. and *Salmonella* spp. combined. Although *C. jejuni* is not carried by healthy individuals in the United States or Europe, it is often isolated from healthy cattle, chickens, birds and even flies. It is sometimes present in non-chlorinated water sources such as streams and ponds. Because the pathogenic mechanisms of *C. jejuni* are still being studied, it is difficult to differentiate pathogenic from nonpathogenic strains. However, it appears that many of the chicken isolates are pathogens.

2. *Name of Disease*: Campylobacteriosis is the name of the illness caused by *C. jejuni*. It is also often known as campylobacter enteritis or gastroenteritis.

3. *Major Symptoms*: *C. jejuni* infection causes diarrhea, which may be watery or sticky and can contain blood (usually occult) and fecal leukocytes (white cells). Other symptoms often present are fever, abdominal pain, nausea, headache and muscle pain. The illness usually occurs 2-5 days after ingestion of the contaminated food or water. Illness generally lasts 7-10 days, but relapses are not uncommon (about 25% of cases). Most infections are self-limiting and are not treated with antibiotics. However, treatment with erythromycin does reduce the length of time that infected individuals shed the bacteria in their feces. The infective dose of *C. jejuni* is considered to be small. Human feeding studies suggest that about 400-500 bacteria may cause illness in some individuals, while in others, greater numbers are required. A conducted volunteer human feeding study suggests that host susceptibility also dictates infectious dose to some degree. The pathogenic mechanisms of *C. jejuni* are still not completely understood, but it does produce a heat-labile toxin that may cause diarrhea. *C. jejuni* may also be an invasive organism.

4. Isolation Procedures: *C. jejuni* is usually present in high numbers in the diarrheal stools of individuals, but isolation requires special antibiotic-containing media and a special microaerophilic atmosphere (5% oxygen). However, most clinical laboratories are equipped to isolate Campylobacter spp. if requested.

5. *Associated Foods*: *C. jejuni* frequently contaminates raw chicken. Surveys show that 20 to 100% of retail chickens are contaminated. This is not overly surprising since many healthy chickens carry these bacteria in their intestinal tracts. Raw milk is also a source of infections. The bacteria are often carried by healthy cattle and by flies on farms. Non-chlorinated water may also be a source of infections. However, properly cooking chicken, pasteurizing milk, and chlorinating drinking water will kill the bacteria.

6. *Frequency of the Disease*: *C. jejuni* is the leading cause of bacterial diarrhea in the U.S. There are probably numbers of cases in excess of the estimated cases of salmonellosis (2- to 4,000,000/year).

7. *Complications*: Complications are relatively rare, but infections have been associated with reactive arthritis, hemolytic uremic syndrome, and following septicemia, infections of nearly any organ. The estimated case/fatality ratio for all *C. jejuni* infections is 0.1, meaning one death per 1,000 cases. Fatalities are rare in healthy individuals and usually occur in cancer patients or in the otherwise debilitated. Only 20 reported cases of septic abortion induced by *C. jejuni* have been recorded in the literature. Meningitis, recurrent colitis, acute cholecystitis and Guillain-Barre syndrome are very rare complications.

8. Target Populations: Although anyone can have a *C. jejuni* infection, children under 5 years and young adults (15-29) are more frequently afflicted than other age groups. Reactive arthritis, a rare complication of these infections, is strongly associated with people who have the human lymphocyte antigen B27 (HLA-B27).

9. Recovery from Foods: Isolation of *C. jejuni* from food is difficult because the bacteria are usually present in very low numbers (unlike the case of diarrheal stools in which 10/6 bacteria/gram is not unusual). The methods require an enrichment broth containing antibiotics, special antibiotic-containing plates and a microaerophilic atmosphere generally a microaerophilic atmosphere with 5% oxygen and an elevated concentration of carbon dioxide (10%). Isolation can take several days to a week.

10. *Selected Outbreaks*: Usually outbreaks are small (less than 50 people), but in Bennington, VT a large outbreak involving about 2,000 people occurred while the town was temporarily using an non-chlorinated water source as a water supply. Several small outbreaks have been reported among children who were taken on a class trip to a dairy and given raw milk to drink. An outbreak was also associated with consumption of raw clams. However, a survey showed that about 50% of infections are associated with either eating inadequately cooked or recontaminated chicken meat or handling chickens.

It is the leading bacterial cause of sporadic (non-clustered cases) diarrheal disease in the U.S. In April, 1986, an elementary school child was cultured for bacterial pathogens (due to bloody diarrhea), and *C. jejuni* was isolated. Food consumption/gastrointestinal illness questionnaires were administered to other students and faculty at the school. In all, 32 of 172 students reported symptoms of diarrhea (100%), cramps (80%), nausea (51%), fever (29%), vomiting (26%), and bloody stools (14%). The food questionnaire clearly implicated milk as the common source, and a dose/response was evident (those drinking more milk were more likely to be ill). Investigation of the dairy supplying the milk showed that they vat pasteurized the milk at 135°F for 25 minutes rather than the required 145°F for 30 minutes. The dairy processed surplus raw milk for the school, and this milk had a high somatic cell count. Cows from the herd supplying the dairy had *C. jejuni* in their feces. This outbreak points out the variation in symptoms which may occur with campylobacteriosis and the absolute need to adhere to pasteurization time/temperature standards.

Yersinia Enterocolitica

1. Name of the Organism: *Yersinia enterocolitica* (and *Yersinia pseudotuberculosis*) *Y. enterocolitica,* a small rod-shaped, Gram- negative bacterium, is often isolated from clinical specimens such as wounds, feces, sputum and mesenteric lymph nodes. However, it is not part of the normal human flora. *Y. pseudotuberculosis* has been isolated from the diseased appendix of humans. Both organisms have often been isolated from such animals as pigs, birds, beavers, cats, and dogs. Only *Y. enterocolitica* has been detected in environmental and food sources, such as ponds, lakes, meats, ice cream, and milk. Most isolates have been found not to be pathogenic.

2. *Name of Disease*: Yersiniosis There are 3 pathogenic species in the genus Yersinia, but only *Y. enterocolitica* and *Y. pseudotuberculosis* cause gastroenteritis. To date, no foodborne outbreaks caused by

Y. pseudotuberculosis have been reported in the United States, but human infections transmitted via contaminated water and foods have been reported in Japan. *Y. pestis,* the causative agent of "the plague," is genetically very similar to *Y. pseudotuberculosis* but infects humans by routes other than food.

3. *Nature of Disease*: Yersiniosis is frequently characterized by such symptoms as gastroenteritis with diarrhea and/or vomiting; however, fever and abdominal pain are the hallmark symptoms. *Yersinia* infections mimic appendicitis and mesenteric lymphadenitis, but the bacteria may also cause infections of other sites such as wounds, joints and the urinary tract.

4. *Infective dose*: Unknown. Illness onset is usually between 24 and 48 hours after ingestion, which (with food or drink as vehicle) is the usual route of infection.

5. *Diagnosis of Human Illness*: Diagnosis of yersiniosis begins with isolation of the organism from the human host's feces, blood, or vomit, and sometimes at the time of appendectomy. Confirmation occurs with the isolation, as well as biochemical and serological identification, of

Y. enterocolitica from both the human host and the ingested foodstuff. Diarrhea is reported to occur in about 80% of cases; abdominal pain and fever are the most reliable symptoms. Because of the difficulties in isolating *yersiniae* from feces, several countries rely on serology. Acute and convalescent patient sera are titered against the suspect serotype of *Yersinia spp*. Yersiniosis has been misdiagnosed as Crohn's disease (regional enteritis) as well as appendicitis.

6. *Associated Foods*: Strains of *Y. enterocolitica* can be found in meats (pork, beef, lamb, etc.), oysters, fish, and raw milk. The exact cause of the food contamination is unknown. However, the prevalence of this organism in the soil and water and in animals such as beavers, pigs, and squirrels, offers ample opportunities for it to enter our food supply. Poor sanitation and improper sterilization techniques by food handlers, including improper storage, cannot be overlooked as contributing to contamination.

7. *Frequency of the Disease*: Yersiniosis does not occur frequently. It is rare unless a breakdown occurs in food processing techniques. CDC estimates that about 17,000 cases occur annually in the USA. Yersiniosis is a far more common disease in Northern Europe, Scandinavia, and Japan.

8. *Complications*: The major "complication" is the performance of unnecessary appendectomies, since one of the main symptoms of infections is abdominal pain of the lower right quadrant. Both *Y. enterocolitica* and *Y. pseudotuberculosis* have been associated with reactive arthritis, which may occur even in the absence of obvious symptoms. The frequency of such postenteritis arthritic conditions is about 2-3%. Another complication is bacteremia (entrance of organisms into the blood stream), in which case the possibility of a disseminating disease may occur. This is rare, however, and fatalities are also extremely rare.

9. *Target Populations*: The most susceptible populations for the main disease and possible complications are the very young, the debilitated, the very old and persons undergoing immunosuppressive therapy. Those most susceptible to postenteritis arthritis are individuals with the antigen HLA-B27 (or related antigens such as B7).

10. Food Analysis: The isolation method is relatively easy to perform, but in some instances, cold enrichment may be required. *Y. enterocolitica* can be presumptively identified in 36-48 hours. However, confirmation may take 14-21 days or more. Determination of pathogenicity is more complex. The genes encoding for invasion of mammalian cells are located on the chromosome while a 40-50 MDal plasmid encodes most of the other virulence associated phenotypes. The 40-50 MDal plasmid is present in almost all the pathogenic *Yersinia* species, and the plasmids appear to be homologous.

11. Selected Outbreaks: 1976. A chocolate milk outbreak in Oneida County, N.Y. involving school children (first reported yersiniosis incident in the United States in which a food vehicle was identified). A research laboratory was set up by FDA to investigate and study

Y. enterocolitica and *Y. pseudotuberculosis* in the human food supply. Dec. 1981 - Feb. 1982. *Y. enterocolitica* enteritis in King County, Washington caused by ingestion of tofu, a soybean curd. FDA investigators and researchers determined the source of the infection to be an non-chlorinated water supply. Manufacturing was halted until uncontaminated product was produced. June 11 to July 21, 1982. *Y. enterocolitica* outbreak in Arkansas, Tennessee, and Mississippi associated with the consumption of pasteurized milk. FDA personnel participated in the investigation, and presumptively identified the infection source to be externally contaminated milk containers.

Plesiomonas Shigelloides

1. Name of the Organism: *Plesiomonas shigelloides* This is a Gram- negative, rod-shaped bacterium which has been isolated from freshwater, freshwater fish, and shellfish and from many types of animals including cattle, goats, swine, cats, dogs, monkeys, vultures, snakes, and toads. Most human *P. shigelloides* infections are suspected to be waterborne. The organism may be present in unsanitary water which has been used as drinking water, recreational water, or water used to rinse foods that are consumed without cooking or heating. The ingested *P. shigelloides* organism does not always cause illness in the host animal but may reside temporarily as a transient, noninfectious member of the intestinal flora. It has been isolated from the stools of patients with diarrhea, but is also sometimes isolated from healthy individuals (0.2-3.2% of population).It cannot yet be considered a definite cause of human disease, although its association with human diarrhea and the virulence factors it demonstrates make it a prime candidate.

2. Nature of Acute Disease: Gastroenteritis is the disease with which *P. shigelloides* has been implicated.

3. Nature of Disease: *P. shigelloides* gastroenteritis is usually a mild self-limiting disease with fever, chills, abdominal pain, nausea, diarrhea, or vomiting; symptoms may begin 20-24 hours after consumption of contaminated food or water; diarrhea is watery, non-mucoid, and non-bloody; in severe cases, diarrhea may be greenish-yellow, foamy, and blood tinged; duration of illness in healthy people may be 1-7 days. The infectious dose is presumed to be quite high, at least greater than one million organisms.

4. Diagnosis of Human Illness: The pathogenesis of *P. shigelloides* infection is not known. The organism is suspected of being toxigenic and invasive. Its significance as an enteric (intestinal) pathogen is presumed because of its predominant isolation from stools of patients with diarrhea. It is identified by common bacteriological analysis, serotyping, and antibiotic sensitivity testing.

5. Associated Foods: Most *P. shigelloides* infections occur in the summer months and correlate with environmental contamination of freshwater (rivers, streams, ponds, etc.). The usual route of transmission of the organism in sporadic or epidemic cases is by ingestion of contaminated water or raw shellfish.

6. Relative Frequency of Disease: ©Most *P. shigelloides* strains associated with human gastrointestinal disease have been from stools of diarrheic patients living in tropical and subtropical areas. Such infections are rarely reported in the U.S. or Europe because of the self-limiting nature of the disease.

7. Course of Disease and Complications: *P. shigelloides* infection may cause diarrhea of 1-2 days duration in healthy adults. However, there may be high fever and chills and protracted dysenteric symptoms in infants and children under 15 years of age. Extra- intestinal complications (septicemia and death) may occur in people who are immunocompromised or seriously ill with cancer, blood disorders, or hepatobiliary disease.

8. Target Populations: All people may be susceptible to infection. Infants, children and chronically ill people are more likely to experience protracted illness and complications.

9. Food Analysis: *P. shigelloides* may be recovered from food and water by methods similar to those used for stool analysis. The keys to recovery in all cases are selective agars which enhance the survival and growth of these bacteria over the growth of the background microflora. Identification following recovery may be completed in 12-24 hours.

10. Selected Outbreaks: Literature references can be found at the links below. – Gastrointestinal illness in healthy people caused by *P. shigelloides* infection may be so mild that they do not seek medical treatment. Its rate of occurrence in the U.S.

is unknown. It may be included in the group of diarrheal diseases "of unknown etiology" which are treated with and respond to broad spectrum antibiotics.

Streptococcus Spp

1. Name of the Organism: *Streptococcus* spp. The genus Streptococcus is comprised of Gram-positive, microaerophilic cocci (round), which are not motile and occur in chains or pairs. The genus is defined by a combination of antigenic, hemolytic, and physiological characteristics into Groups A, B, C, D, F, and G. Groups A and D can be transmitted to humans via food. Group A: one species with 40 antigenic types (*S. pyogenes*).Group D: five species (*S. faecalis, S. faecium, S. durans, S. avium,* and *S. bovis*).

2. Nature of Acute Disease: Group A: Cause septic sore throat and scarlet fever as well as other pyogenic and septicemic infections. Group D: May produce a clinical syndrome similar to staphylococcal intoxication.

3. Nature of Disease: Group A: Sore and red throat, pain on swallowing, tonsilitis, high fever, headache, nausea, vomiting, malaise, rhinorrhea; occasionally a rash occurs, onset 1-3 days; the infectious dose is probably quite low (less than 1,000 organisms). Group D: Diarrhea, abdominal cramps, nausea, vomiting, fever, chills, dizziness in 2-36 hours. Following ingestion of suspect food, the infectious dose is probably high (greater than 107 organisms).

4. Diagnosis of Human Illness: Group A: Culturing of nasal and throat swabs, pus, sputum, blood, suspect food, environmental samples. Group D: Culturing of stool samples, blood, and suspect food.

5. Associated Foods: Group A: Food sources include milk, ice cream, eggs, steamed lobster, ground ham, potato salad, egg salad, custard, rice pudding, and shrimp salad. In almost all cases, the foodstuffs were allowed to stand at room temperature for several hours between preparation and consumption. Entrance into the food is the result of poor hygiene, ill food handlers, or the use of unpasteurized milk. Group D: Food sources include sausage, evaporated milk, cheese, meat croquettes, meat pie, pudding, raw milk, and pasteurized milk. Entrance into the food chain is due to underprocessing and/or poor and unsanitary food preparation.

6. Relative Frequency of Disease: Group A infections are low and may occur in any season, whereas Group D infections are variable.

7. Course of Disease and Complications: Group A: Streptococcal sore throat is very common, especially in children. Usually it is successfully treated with antibiotics. Complications are rare and the fatality rate is low. Group D: Diarrheal illness is poorly characterized, but is acute and self-limiting.

8. *Target Populations*: All individuals are susceptible. No age or race susceptibilities have been found.

9. *Food Analysis*: Suspect food is examined microbiologically by selective enumeration techniques which can take up to 7 days. Group specificities are determined by Lancefield group-specific antisera.

10. *Selected Outbreaks: Literature references can be found at the links below.*

- Group A: Outbreaks of septic sore throat and scarlet fever were numerous before the advent of milk pasteurization. Salad bars have been suggested as possible sources of infection. Most current outbreaks have involved complex foods (i.e., salads) which were infected by a food handler with septic sore throat. One ill food handler may subsequently infect hundreds of individuals.
- Group D: Outbreaks are not common and are usually the result of preparing, storing, or handling food in an unsanitary manner.

NON-BACTERIAL CAUSES

PROTOZOA AND WORMS

Giardia Lamblia

1. *Name of the Organism*: *Giardia lamblia Giardia lamblia* (intestinalis) is a single celled animal, i.e., a protozoa, that moves with the aid of five flagella. In Europe, it is sometimes referred to as Lamblia intestinalis.

2. *Nature of Acute Disease*: Giardiasis is the most frequent cause of non-bacterial diarrhea in North America.

3. *Nature of Disease*: Organisms that appear identical to those that cause human illness have been isolated from domestic animals (dogs and cats) and wild animals (beavers and bears). A related but morphologically distinct organism infects rodents, although rodents may be infected with human isolates in the laboratory. Human giardiasis may involve diarrhea within 1 week of ingestion of the cyst, which is the environmental survival form and infective stage of the organism. Normally illness lasts for 1 to 2 weeks, but there are cases of chronic infections lasting months to years.

Chronic cases, both those with defined immune deficiencies and those without, are difficult to treat.The disease mechanism is unknown, with some investigators reporting that the organism produces a toxin while others are unable to confirm its existence. The organism has been demonstrated inside host cells in the duodenum, but most investigators think this is such an infrequent occurrence that it is not responsible for disease symptoms.

Mechanical obstruction of the absorptive surface of the intestine has been proposed as a possible pathogenic mechanism, as has a synergistic relationship with some of the intestinal flora.Giardia can be excysted, cultured and encysted in vitro; new isolates have bacterial, fungal, and viral symbionts. Classically the disease was diagnosed by demonstration of the organism in stained fecal smears.Several strains of *G. lamblia* have been isolated and described through analysis of their proteins and DNA; type of strain, however, is not consistently associated with disease severity. Different individuals show various degrees of symptoms when infected with the same strain, and the symptoms of an individual may vary during the course of the disease.Infectious Dose - Ingestion of one or more cysts may cause disease, as contrasted to most bacterial illnesses where hundreds to thousands of organisms must be consumed to produce illness.

4. *Diagnosis of Human Illness*: *Giardia lamblia* is frequently diagnosed by visualizing the organism, either the trophozoite (active reproducing form) or the cyst (the resting stage that is resistant to adverse environmental conditions) in stained preparations or unstained wet mounts with the aid of a microscope. A commercial fluorescent antibody kit is available to stain the organism. Organisms may be concentrated by sedimentation or flotation; however, these procedures reduce the number of recognizable organisms in the sample. An enzyme linked immunosorbant assay (ELISA) that detects excretory secretory products of the organism is also available. So far, the increased sensitivity of indirect serological detection has not been consistently demonstrated.

5. *Associated Foods*: Giardiasis is most frequently associated with the consumption of contaminated water. Five outbreaks have been traced to food contamination by infected or infested food handlers, and the possibility of infections from contaminated vegetables that are eaten raw cannot be excluded. Cool moist conditions favor the survival of the organism.

6. *Relative Frequency of Disease*: Giardiasis is more prevalent in children than in adults, possibly because many individuals seem to have a lasting immunity after infection. This organism is implicated in 25% of the cases of gastrointestinal disease and may be present asymptomatically. The overall incidence of infection in the United States is estimated at 2% of the population. This disease afflicts many homosexual men, both HIV-positive and HIV-negative individuals. This is presumed to be due to sexual transmission. The disease is also common in child day care centers, especially those in which diapering is done.

7. *Course of Disease and Complications*: About 40% of those who are diagnosed with giardiasis demonstrate disaccharide intolerance during

detectable infection and up to 6 months after the infection can no longer be detected. Lactose (i.e., milk sugar) intolerance is most frequently observed. Some individuals (less than 4%) remain symptomatic more than 2 weeks; chronic infections lead to a malabsorption syndrome and severe weight loss. Chronic cases of giardiasis in immunodeficient and normal individuals are frequently refractile to drug treatment. Flagyl is normally quite effective in terminating infections. In some immune deficient individuals, giardiasis may contribute to a shortening of the life span.

8. *Target Populations*: Giardiasis occurs throughout the population, although the prevalence is higher in children than adults. Chronic symptomatic giardiasis is more common in adults than children.

9. *Food Analysis*: Food is analyzed by thorough surface cleaning of the suspected food and sedimentation of the organisms from the cleaning water. Feeding to specific pathogen-free animals has been used to detect the organism in large outbreaks associated with municipal water systems. The precise sensitivity of these methods has not been determined, so that negative results are questionable. Seven days may be required to detect an experimental infection.

Entamoeba Histolytica

1. *Name of the Organism*: *Entamoeba histolytica* This is a single celled parasitic animal, i.e., a protozoa, that infects predominantly humans and other primates. Diverse mammals such as dogs and cats can become infected but usually do not shed cysts (the environmental survival form of the organism) with their feces, thus do not contribute significantly to transmission. The active (trophozoite) stage exists only in the host and in fresh feces; cysts survive outside the host in water and soils and on foods, especially under moist conditions on the latter. When swallowed they cause infections by excysting (to the trophozoite stage) in the digestive tract.

2. *Nature of Acute Disease*: Amebiasis (or amoebiasis) is the name of the infection caused by *E. histolytica.*

3. *Nature of Disease*: Infections that sometimes last for years may be accompanied by 1) no symptoms, 2) vague gastrointestinal distress, 3) dysentery (with blood and mucus). Most infections occur in the digestive tract but other tissues may be invaded. Complications include

4) ulcerative and abscess pain and, rarely, 5) intestinal blockage. Onset time is highly variable. It is theorized that the absence of symptoms or their intensity varies with such factors as 1) strain of amoeba, 2) immune health of the host, and 3) associated bacteria and, perhaps, viruses. The

amoeba's enzymes help it to penetrate and digest human tissues; it secretes toxic substances. Infectious Dose—Theoretically, the ingestion of one viable cyst can cause an infection.

4. *Diagnosis of Human Illness*: Human cases are diagnosed by finding cysts shed with the stool; various flotation or sedimentation procedures have been developed to recover the cysts from fecal matter; stains (including fluorescent antibody) help to visualize the isolated cysts for microscopic examination. Since cysts are not shed constantly, a minimum of 3 stools should be examined. In heavy infections, the motile form (the trophozoite) can be seen in fresh feces. Serological tests exist for long-term infections. It is important to distinguish the *E. histolytica* cyst from the cysts of nonpathogenic intestinal protozoa by its appearance.

5. *Associated Foods*: Amebiasis is transmitted by fecal contamination of drinking water and foods, but also by direct contact with dirty hands or objects as well as by sexual contact.

6. *Relative Frequency of Disease*: The infection is "not uncommon" in the tropics and arctics, but also in crowded situations of poor hygiene in temperate-zone urban environments. It is also frequently diagnosed among homosexual men.

7. *Course of Disease and Complications*: In the majority of cases, amoebas remain in the gastrointestinal tract of the hosts. Severe ulceration of the gastrointestinal mucosal surfaces occurs in less than 16% of cases. In fewer cases, the parasite invades the soft tissues, most commonly the liver. Only rarely are masses formed (amoebomas) that lead to intestinal obstruction. Fatalities are infrequent.

8. *Target Populations*: All people are believed to be susceptible to infection, but individuals with a damaged or undeveloped immunity may suffer more severe forms of the disease. AIDS / ARC patients are very vulnerable.

9. *Food Analysis*: *E. histolytica* cysts may be recovered from contaminated food by methods similar to those used for recovering *Giardia lamblia* cysts from feces. Filtration is probably the most practical method for recovery from drinking water and liquid foods.

E. histolytica cysts must be distinguished from cysts of other parasitic (but nonpathogenic) protozoa and from cysts of free-living protozoa. Recovery procedures are not very accurate; cysts are easily lost or damaged beyond recognition, which leads to many falsely negative results in recovery tests.

Ascaris Lumbricoides and Trichuris Trichiura

1. *Name of the Organism*: *Ascaris lumbricoides* and *Trichuris trichiura* Humans worldwide are infected with *Ascaris lumbricoides* and *Trichuris trichiura*; the eggs

of these roundworms (nematode) are "sticky" and may be carried to the mouth by hands, other body parts, fomites (inanimate objects), or foods.

2. Nature of Acute Disease: Ascariasis and trichuriasis are the scientific names of these infections. Ascariasis is also known commonly as the "large roundworm" infection and trichuriasis as "whip worm" infection.

3. *Nature of Disease*: Infection with one or a few *Ascaris* sp. may be inapparent unless noticed when passed in the feces, or, on occasion, crawling up into the throat and trying to exit through the mouth or nose. Infection with numerous worms may result in a pneumonitis during the migratory phase when larvae that have hatched from the ingested eggs in the lumen of the small intestine penetrate into the tissues and by way of the lymph and blood systems reach the lungs. In the lungs, the larvae break out of the pulmonary capillaries into the air sacs, ascend into the throat and descend to the small intestine again where they grow, becoming as large as 31 X 4 cm. Molting (ecdysis) occurs at various points along this path and, typically for roundworms, the male and female adults in the intestine are 5th- stage nematodes. Vague digestive tract discomfort sometimes accompanies the intestinal infection, but in small children with more than a few worms there may be intestinal blockage because of the worms' large size. Not all larval or adult worms stay on the path that is optimal for their development; those that wander may locate in diverse sites throughout the body and cause complications. Chemotherapy with anthelmintics is particularly likely to cause the adult worms in the intestinal lumen to wander; a not unusual escape route for them is into the bile duct which they may occlude. The larvae of ascarid species that mature in hosts other than humans may hatch in the human intestine and are especially prone to wander; they may penetrate into tissues and locate in various organ systems of the human body, perhaps eliciting a fever and diverse complications. *Trichuris* sp. larvae do not migrate after hatching but molt and mature in the intestine. Adults are not as large as *A. lumbricoides*. Symptoms range from inapparent through vague digestive tract distress to emaciation with dry skin and diarrhea (usually mucoid). Toxic or allergic symptoms may also occur.

4. *Diagnosis of Human Illness*: Both infections are diagnosed by finding the typical eggs in the patient's feces; on occasion the larval or adult worms are found in the feces or, especially for *Ascaris* sp., in the throat, mouth, or nose.

5. *Associated Foods*: The eggs of these worms are found in insufficiently treated sewage-fertilizer and in soils where they embryonate (i.e., larvae develop in fertilized eggs). The eggs may contaminate crops grown in soil

or fertilized with sewage that has received nonlethal treatment; humans are infected when such produce is consumed raw. Infected foodhandlers may contaminate a wide variety of foods.

6. *Relative Frequency of Disease:* These infections are cosmopolitan, but ascariasis is more common in North America and trichuriasis in Europe. Relative infection rates on other continents are not available.

7. *Course of Disease and Complications*: Both infections may self-cure after the larvae have matured into adults or may require anthelmintic treatment. In severe cases, surgical removal may be necessary. Allergic symptoms (especially but not exclusively of the asthmatic sort) are common in long-lasting infections or upon reinfection in ascariasis.

8. *Target Populations*: Particularly consumers of uncooked vegetables and fruits grown in or near soil fertilized with sewage.

9. *Food Analysis:* Eggs of *Ascaris* spp. have been detected on fresh vegetables (cabbage) sampled by FDA

Viruses

Hepatitis A Virus

1. *Name of the Organism*: Hepatitis A Virus Hepatitis A virus (HAV) is classified with the enterovirus group of the Picornaviridae family. HAV has a single molecule of RNA surrounded by a small (27 nm diameter) protein capsid and a buoyant density in CsCl of 1.33 g/ ml. Many other picornaviruses cause human disease, including polioviruses, coxsackieviruses, echoviruses, and rhinoviruses (cold viruses).

2. *Nature of Acute Disease*: The term hepatitis A (HA) or type A viral hepatitis has replaced all previous designations: infectious hepatitis, epidemic hepatitis, epidemic jaundice, catarrhal jaundice, infectious icterus, Botkins disease, and MS-1 hepatitis.

3. *Nature of Disease*: Hepatitis A is usually a mild illness characterized by sudden onset of fever, malaise, nausea, anorexia, and abdominal discomfort, followed in several days by jaundice. The infectious dose is unknown but presumably is 10-100 virus particles.

4. *Diagnosis of Human Illness*: Hepatitis A is diagnosed by finding IgM-class anti-HAV in serum collected during the acute or early convalescent phase of disease. Commercial kits are available.

5. Associated Foods: HAV is excreted in feces of infected people and can produce clinical disease when susceptible individuals consume contaminated water or foods. Cold cuts and sandwiches, fruits and fruit juices, milk and

milk products, vegetables, salads, shellfish, and iced drinks are commonly implicated in outbreaks. Water, shellfish, and salads are the most frequent sources. Contamination of foods by infected workers in food processing plants and restaurants is common.

6. *Relative Frequency of Disease*: Hepatitis A has a worldwide distribution occurring in both epidemic and sporadic fashions. About 22,700 cases of hepatitis A representing 38% of all hepatitis cases (5-year average from all routes of transmission) are reported annually in the U.S. In 1988 an estimated 7.3% cases were foodborne or waterborne. HAV is primarilly transmitted by person-to-person contact through fecal contamination, but common-source epidemics from contaminated food and water also occur. Poor sanitation and crowding facilitate transmission. Outbreaks of HA are common in institutions, crowded house projects, and prisons and in military forces in adverse situations. In developing countries, the incidence of disease in adults is relatively low because of exposure to the virus in childhood. Most individuals 18 and older demonstrate an immunity that provides lifelong protection against reinfection. In the U.S., the percentage of adults with immunity increases with age (10% for those 18-19 years of age to 65% for those over 50). The increased number of susceptible individuals allows common source epidemics to evolve rapidly.

7. *Course of Disease and Complications*: The incubation period for hepatitis A, which varies from 10 to 50 days (mean 30 days), is dependent upon the number of infectious particles consumed. Infection with very few particles results in longer incubation periods. The period of communicability extends from early in the incubation period to about a week after the development of jaundice. The greatest danger of spreading the disease to others occurs during the middle of the incubation period, well before the first presentation of symptoms.

Many infections with HAV do not result in clinical disease, especially in children. When disease does occur, it is usually mild and recovery is complete in 1-2 weeks. Occasionaly, the symptoms are severe and convalescence can take several months. Patients suffer from feeling chronically tired during convalescence, and their inability to work can cause financial loss. Less than 0.4% of the reported cases in the U.S. are fatal. These rare deaths usually occur in the elderly.

8. *Target Populations*: All people who ingest the virus and are immunologically unprotected are susceptible to infection. Disease however, is more common in adults than in children.

9. *Food Analysis*: The virus has not been isolated from any food associated with an outbreak. Because of the long incubation period, the suspected food is often no longer available for analysis. No satisfactory method is presently available for routine analysis of food, but sensitive molecular methods used to detect HAV in water and clinical specimens, should prove useful to detect virus in foods. Among those, the PCR amplification method seems particularly promising.

Hepatitis E Virus

1. *Name of the Organism*: Hepatitis E Virus Hepatitis E Virus (HEV) has a particle diameter of 32-34 nm, a buoyant density of 1.29 g/ ml in KTar/Gly gradient, and is very labile. Serologically related smaller (27-30 nm) particles are often found in feces of patients with Hepatitis E and are presumed to represent degraded viral particles. HEV has a single-stranded polyadenylated RNA genome of approximately 8 kb. Based on its physicochemical properties it is presumed to be a calici-like virus.

2. *Nature of Acute Disease*: The disease caused by HEV is called hepatitis E, or enterically transmitted non-A non-B hepatitis (ET-NANBH). Other names include fecal-oral non-A non-B hepatitis,and A-like non-A non-B hepatitis. Note: This disease should not be confused with hepatitis C, also called parenterally transmitted non-A non-B hepatitis (PT-NANBH), or B-like non-A non-B hepatitis, which is a common cause of hepatitis in the U.S.

3. *Nature of Disease*: Hepatitis caused by HEV is clinically indistinguishable from hepatitis A disease. Symptoms include malaise, anorexia, abdominal pain, arthralgia, and fever. The infective dose is not known.

4. Diagnosis of Human Illness: Diagnosis of HEV is based on the epidemiological characteristics of the outbreak and by exclusion of hepatitis A and B viruses by serological tests. Confirmation requires identification of the 27-34 nm virus-like particles by immune electron microscopy in feces of acutely ill patients.

5. *Associated Foods*: HEV is transmitted by the fecal-oral route. Waterborne and person-to-person spread have been documented. The potential exists for foodborne transmission.

6. *Relative Frequency of Disease*: Hepatitis E occurs in both epidemic and sporadic-endemic forms, usually associated with contaminated drinking water. Major waterborne epidemics have occurred in Asia and North and East Africa. To date no U.S. outbreaks have been reported.

7. *Course of Disease and Complications*: The incubation period for hepatitis E varies from 2 to 9 weeks. The disease usually is mild and resolves in 2

weeks, leaving no sequelae. The fatality rate is 0.1-1% except in pregnant women. This group is reported to have a fatality rate approaching 20%.

8. *Target Populations*: The disease is most often seen in young to middle aged adults (15-40 years old). Pregnant women appear to be exceptionally susceptible to severe disease, and excessive mortality has been reported in this group.

9. *Food Analysis*: HEV has not been isolated from foods. No method is currently available for routine analysis of foods.

Rotavirus

1. *Name of the Organism*: Rotavirus Rotaviruses are classified with the Reoviridae family. They have a genome consisting of 11 double- stranded RNA segments surrounded by a distinctive two-layered protein capsid. Particles are 70 nm in diameter and have a buoyant density of 1.36 g/ml in CsCl. Six serological groups have been identified, three of which (groups A, B, and C) infect humans.

2. *Nature of Acute Disease*: Rotaviruses cause acute gastroenteritis. Infantile diarrhea, winter diarrhea, acute nonbacterial infectious gastroenteritis, and acute viral gastroenteritis are names applied to the infection caused by the most common and widespread group A rotavirus.

3. *Nature of Disease*: Rotavirus gastroenteritis is a self-limiting, mild to severe disease characterized by vomiting, watery diarrhea, and low- grade fever. The infective dose is presumed to be 10-100 infectious viral particles. Because a person with rotavirus diarrhea often excretes large numbers of virus (108-1010 infectious particles/ml of feces), infection doses can be readily acquired through contaminated hands, objects, or utensils. Asymptomatic rotavirus excretion has been well documented and may play a role in perpetuating endemic disease.

4. *Diagnosis of Human Illness*: Specific diagnosis of the disease is made by identification of the virus in the patient's stool. Enzyme immunoassay (EIA) is the test most widely used to screen clinical specimens, and several commercial kits are available for group A rotavirus. Electron microscopy (EM) and polyacrylamide gel electrophoresis (PAGE) are used in some laboratories in addition or as an alternative to EIA. A reverse transcription-polymerase chain reaction (RT-PCR) has been developed to detect and identify all three groups of human rotaviruses.

5. *Associated Foods*: Rotaviruses are transmitted by the fecal-oral route. Person-to-person spread through contaminated hands is probably the most important means by which rotaviruses are transmitted in close communities

such as pediatric and geriatric wards, day care centers and family homes. Infected food handlers may contaminate foods that require handling and no further cooking, such as salads, fruits, and hors d'oeuvres. Rotaviruses are quite stable in the environment and have been found in estuary samples at levels as high as 1-5 infectious particles/gal. Sanitary measures adequate for bacteria and parasites seem to be ineffective in endemic control of rotavirus, as similar incidence of rotavirus infection is observed in countries with both high and low health standards.

6. *Relative Frequency of Disease*: Group A rotavirus is endemic worldwide. It is the leading cause of severe diarrhea among infants and children, and accounts for about half of the cases requiring hospitalization. Over 3 million cases of rotavirus gastroenteritis occur annually in the U.S. In temperate areas, it occurs primarily in the winter, but in the tropics it occurs throughout the year. The number attributable to food contamination is unknown. Group B rotavirus, also called adult diarrhea rotavirus or ADRV, has caused major epidemics of severe diarrhea affecting thousands of persons of all ages in China.Group C rotavirus has been associated with rare and sporadic cases of diarrhea in children in many countries. However, the first outbreaks were reported from Japan and England.

7. *Course of Disease and Complications*: The incubation period ranges from 1-3 days. Symptoms often start with vomiting followed by 4- 8 days of diarrhea. Temporary lactose intolerance may occur. Recovery is usually complete. However, severe diarrhea without fluid and electrolyte replacement may result in severe diarrhea and death. Childhood mortality caused by rotavirus is relatively low in the U.S., with an estimated 100 cases/year, but reaches almost 1 million cases/year worldwide. Association with other enteric pathogens may play a role in the severity of the disease.

8. *Target Populations*: Humans of all ages are susceptible to rotavirus infection. Children 6 months to 2 years of age, premature infants, the elderly, and the immunocompromised are particularly prone to more severe symptoms caused by infection with group A rotavirus.

9. *Food Analysis*: The virus has not been isolated from any food associated with an outbreak, and no satisfactory method is available for routine analysis of food. However, it should be possible to apply procedures that have been used to detect the virus in water and in clinical specimens, such as enzyme immunoassays, gene probing, and PCR amplification to food analysis.

The Norwalk Virus Family

1. *Name of the Organism*: The Norwalk virus family Norwalk virus is

the prototype of a family of unclassified small round structured viruses (SRSVs) which may be related to the caliciviruses. They contain a positive strand RNA genome of 7.5 kb and a single structural protein of about 60 kDa. The 27-32 nm viral particles have a buoyant density of 1.39-1.40 g/ml in CsCl. The family consists of several serologically distinct groups of viruses that have been named after the places where the outbreaks occurred. In the U.S., the Norwalk and Montgomery County agents are serologically related but distinct from the Hawaii and Snow Mountain agents. The Taunton, Moorcroft, Barnett, and Amulree agents were identified in the U.K., and the Sapporo and Otofuke agents in Japan. Their serological relationships remain to be determined.

2. *Nature of Acute Disease*: Common names of the illness caused by the Norwalk and Norwalk-like viruses are viral gastroenteritis, acute nonbacterial gastroenteritis, food poisoning, and food infection.

3. *Nature of Disease*: The disease is self-limiting, mild, and characterized by nausea, vomiting, diarrhea, and abdominal pain. Headache and low-grade fever may occur. The infectious dose is unknown but presumed to be low.

4. *Diagnosis of Human Illness*: Specific diagnosis of the disease can only be made by a few laboratories possessing reagents from human volunteer studies. Identification of the virus can be made on early stool specimens using immune electron microscopy and various immunoassays. Confirmation often requires demonstration of seroconversion, the presence of specific IgM antibody, or a four- fold rise in antibody titer to Norwalk virus on paired acute- convalescent sera.

5. *Associated Foods*: Norwalk gastroenteritis is transmitted by the fecal-oral route via contaminated water and foods. Secondary person-to- person transmission has been documented. Water is the most common source of outbreaks and may include water from municipal supplies, well, recreational lakes, swiming pools, and water stored aboard cruise ships. Shellfish and salad ingredients are the foods most often implicated in Norwalk outbreaks. Ingestion of raw or insufficiently steamed clams and oysters poses a high risk for infection with Norwalk virus. Foods other than shellfish are contaminated by ill food handlers.

6. *Relative Frequency of Disease*: Only the common cold is reported more frequently than viral gastroenteritis as a cause of illness in the U.S. Although viral gastroenteritis is caused by a number of viruses, it is estimated that Norwalk viruses are responsible for about 1/3 of the cases not involving the 6-to-24-month age group. In developing countries the percentage of individuals who have developed immunity is very high at an early age. In

the U.S. the percentage increases gradually with age, reaching 50% in the population over 18 years of age. Immunity, however, is not permanent and reinfection can occur.

7. Course of Disease and Complications: A mild and brief illness usually develops 24-48 h after contaminated food or water is consumed and lasts for 24-60 hours. Severe illness or hospitalization is very rare.

8. Target Populations: All individuals who ingest the virus and who have not (within 24 months) had an infection with the same or related strain, are susceptible to infection and can develop the symptoms of gastroenteritis. Disease is more frequent in adults and older children than in the very young.

9. Food Analysis: The virus has been identified in clams and oysters by radioimmunoassay. The genome of Norwalk virus has been cloned and development of gene probes and PCR amplification techniques to detect the virus in clinical specimens and possibly in food are under way.

Natural Toxins Ciguatera

1. Name of the Organism: Ciguatera

2. Nature of Acute Disease: Ciguatera Fish Poisoning Ciguatera is a form of human poisoning caused by the consumption of subtropical and tropical marine finfish which have accumulated naturally occurring toxins through their diet. The toxins are known to originate from several dinoflagellate (algae) species that are common to ciguatera endemic regions in the lower latitudes.

3. Nature of Disease: Manifestations of ciguatera in humans usually involves a combination of gastrointestinal, neurological, and cardiovascular disorders. Symptoms defined within these general categories vary with the geographic origin of toxic fish.

4. Diagnosis of Human Illness: Clinical testing procedures are not presently available for the diagnosis of ciguatera in humans. Diagnosis is based entirely on symptomology and recent dietary history. An enzyme immunoassay (EIA) designed to detect toxic fish in field situations is under evaluation by the Association of Official Analytical Chemists (AOAC) and may provide some measure of protection to the public in the future.

5. Associated Foods: Marine finfish most commonly implicated in ciguatera fish poisoning include the groupers, barracudas, snappers, jacks, mackerel, and triggerfish. Many other species of warm-water fishes harbor ciguatera toxins. The occurrence of toxic fish is sporadic, and not all fish of a given species or from a given locality will be toxic.

6. Relative Frequency of Disease: The relative frequency of ciguatera fish poisoning in the United States is not known. The disease has only recently become known to the general medical community, and there is a concern that incidence is largely under-reported because of the generally non-fatal nature and short duration of the disease.

7. Course of Disease and Complications: Initial signs of poisoning occur within six hours after consumption of toxic fish and include perioral numbness and tingling (paresthesia), which may spread to the extremities, nausea, vomiting, and diarrhea. Neurological signs include intensified paresthesia, arthralgia, myalgia, headache, temperature sensory reversal and acute sensitivity to temperature extremes, vertigo, and muscular weakness to the point of prostration. Cardiovascular signs include arrhythmia, bradycardia or tachycardia, and reduced blood pressure. Ciguatera poisoning is usually self-limiting, and signs of poisoning often subside within several days from onset. However, in severe cases the neurological symptoms are known to persist from weeks to months. In a few isolated cases neurological symptoms have persisted for several years, and in other cases recovered patients have experienced recurrence of neurological symptoms months to years after recovery. Such relapses are most often associated with changes in dietary habits or with consumption of alcohol. There is a low incidence of death resulting from respiratory and cardiovascular failure.

8. Target Populations: All humans are believed to be susceptible to ciguatera toxins. Populations in tropical/subtropical regions are most likely to be affected because of the frequency of exposure to toxic fishes. However, the increasing per capita consumption of fishery products coupled with an increase in interregional transportation of seafood products has expanded the geographic range of human poisonings.

9. Food Analysis: The ciguatera toxins can be recovered from toxic fish through tedious extraction and purification procedures. The mouse bioassay is a generally accepted method of establishing toxicity of suspect fish. A much simplified EIA method intended to supplant the mouse bioassay for identifying ciguatera toxins is under evaluation.

Various Shellfish-Associated Toxins

1. Name of the Organism: Various Shellfish-Associated Shellfish poisoning is caused by a group of toxins elaborated by planktonic algae (dinoflagellates, in most cases) upon which the shellfish feed. The toxins are accumulated and sometimes metabolized by the shellfish. The 20 toxins responsible for paralytic shellfish poisonings (PSP) are all derivatives of saxitoxin. Diarrheic shellfish poisoning (DSP) is presumably caused by a group of high molecular weight polyethers, including okadaic acid, the dinophysis toxins, the pectenotoxins, and yessotoxin. Neurotoxic shellfish poisoning (NSP) is the result of exposure to a group of polyethers called brevetoxins. Amnesic shellfish poisoning (ASP) is caused by the unusual amino acid, domoic acid, as the contaminant of shellfish.

2. Nature of Acute Disease: Types of Shellfish Poisoning.

- Paralytic Shellfish Poisoning (PSP)
- Diarrheic Shellfish Poisoning (DSP)
- Neurotoxic Shellfish Poisoning (NSP)
- Amnesic Shellfish Poisoning (ASP)

3. *Nature of Disease*: Ingestion of contaminated shellfish results in a wide variety of symptoms, depending upon the toxins(s) present, their concentrations in the shellfish and the amount of contaminated shellfish consumed. In the case of PSP, the effects are predominantly neurological and include tingling, burning, numbness, drowsiness, incoherent speech, and respiratory paralysis. Less well characterized are the symptoms associated with DSP, NSP, and ASP. DSP is primarily observed as a generally mild gastrointestinal disorder, i.e., nausea, vomiting, diarrhea, and abdominal pain accompanied by chills, headache, and fever. Both gastrointestinal and neurological symptoms characterize NSP, including tingling and numbness of lips, tongue, and throat, muscular aches, dizziness, reversal of the sensations of hot and cold, diarrhea, and vomiting. ASP is characterized by gastrointestinal disorders (vomiting, diarrhea, abdominal pain) and neurological problems (confusion, memory loss, disorientation, seizure, coma).

4. *Diagnosis of Human Illness*: Diagnosis of shellfish poisoning is based entirely on observed symptomatology and recent dietary history.

5. *Associated Foods*: All shellfish (filter-feeding molluscs) are potentially toxic. However, PSP is generally associated with mussels, clams, cockles, and scallops; NSP with shellfish harvested along the Florida coast and the Gulf of Mexico; DSP with mussels, oysters, and scallops, and ASP with mussels.

6. *Relative Frequency of Disease*: Good statistical data on the occurrence and severity of shellfish poisoning are largely unavailable, which undoubtedly reflects the inability to measure the true incidence of the disease. Cases are frequently misdiagnosed and, in general, infrequently reported. Of these toxicoses, the most serious from a public health perspective appears to be PSP. The extreme potency of the PSP toxins has, in the past, resulted in an unusually high mortality rate.

7. *Course of Disease and Complications*: PSP: Symptoms of the disease develop fairly rapidly, within 0.5 to 2 hours after ingestion of the shellfish, depending on the amount of toxin consumed. In severe cases respiratory paralysis is common, and death may occur if respiratory support is not provided. When such support is applied within 12 hours of exposure, recovery usually is complete, with no lasting side effects. In unusual cases, because of the weak hypotensive action of the toxin, death may occur from cardiovascular collapse despite respiratory support. NSP: Onset of this disease occurs within a few minutes to a few hours; duration is fairly short, from a few hours to several days. Recovery is complete with few after effects; no fatalities have been reported.DSP: Onset of the disease, depending on the dose of toxin ingested, may be as little as 30 minutes to 2 to 3 hours, with symptoms of the illness lasting as long as 2 to 3 days. Recovery is complete with no after effects; the disease is generally not life threatening.ASP: The toxicosis is characterized by the onset of gastrointestinal symptoms within 24 hours; neurological symptoms occur within 48 hours. The toxicosis is particularly serious in elderly patients, and includes symptoms reminiscent of Alzheimer's disease. All fatalities to date have involved elderly patients.

8. *Target Populations*: All humans are susceptible to shellfish poisoning. Elderly people are apparently predisposed to the severe neurological effects of the ASP toxin. A disproportionate number of PSP cases occur among tourists or others who are not native to the location where the toxic shellfish are harvested. This may be due to disregard for either official quarantines or traditions of safe consumption, both of which tend to protect the local population.

9. *Food Analysis*: The mouse bioassay has historically been the most universally applied technique for examining shellfish (especially for PSP); other bioassay procedures have been developed but not generally applied. Unfortunately, the dose-survival times for the DSP toxins in the mouse assay fluctuate considerably and fatty acids interfere with the assay, giving false-positive results; consequently, a suckling mouse assay that has been developed and used for control of DSP measures fluid accumulation after injection of the shellfish extract. In recent years considerable effort has been applied to development of chemical assays to replace these bioassays. As a result a good high performance liquid chromatography (HPLC) procedure has been developed to identify individual PSP toxins (detection limit for saxitoxin = 20 fg/100 g of meats; 0.2 ppm), an excellent HPLC procedure (detection limit for okadaic acid = 400 ng/g; 0.4 ppm), a commercially available immunoassay (detection limit for okadaic acid = 1 fg/100 g of meats; 0.01 ppm) for DSP and a totally satisfactory HPLC procedure for ASP (detection limit for domoic acid = 750 ng/g; 0.75 ppm).

Scombrotoxin

1. *Name of the Organism*: Scombrotoxin

2. *Nature of Acute Disease*: Scombroid Poisoning (also called Histamine Poisoning) Scombroid poisoning is caused by the ingestion of foods that contain high levels of histamine and possibly other vasoactive amines and compounds. Histamine and other amines are formed by the growth of certain bacteria and the subsequent action of their decarboxylase enzymes on histidine and other amino acids in food, either during the production of a product such as Swiss cheese or by spoilage of foods such as fishery products, particularly tuna or mahi mahi. However, any food that contains the appropriate amino acids and is subjected to certain bacterial contamination and growth may lead to scombroid poisoning when ingested.

3. *Nature of Disease*: Initial symptoms may include a tingling or burning sensation in the mouth, a rash on the upper body and a drop in blood pressure. Frequently, headaches and itching of the skin are encountered. The symptoms may progress to nausea, vomiting, and diarrhea and may require hospitalization, particularly in the case of elderly or impaired patients.

4. *Diagnosis of Human Illness*: Diagnosis of the illness is usually based on the patient's symptoms, time of onset, and the effect of treatment with antihistamine medication. The suspected food must be analyzed within a few hours for elevated levels of histamine to confirm a diagnosis.

5. *Associated Foods*: Fishery products that have been implicated in scombroid poisoning include the tunas (e.g., skipjack and yellowfin), mahi mahi, bluefish, sardines, mackerel, amberjack, and abalone. Many other products also have caused the toxic effects. The primary cheese involved in intoxications has been Swiss cheese. The toxin forms in a food when certain bacteria are present and time and temperature permit their growth. Distribution of the toxin within an individual fish fillet or between cans in a case lot can be uneven, with some sections of a product causing illnesses and others not.

Neither cooking, canning, or freezing reduces the toxic effect. Common sensory examination by the consumer cannot ensure the absence or presence of the toxin. Chemical testing is the only reliable test for evaluation of a product.

6. *Relative Frequency of Disease*: Scombroid poisoning remains one of the most common forms of fish poisoning in the United States. Even so, incidents of poisoning often go unreported because of the lack of required reporting, a lack of information by some medical personnel, and confusion with the symptoms of other illnesses. Difficulties with underreporting are a worldwide

problem. In the United States from 1968 to 1980, 103 incidents of intoxication involving 827 people were reported. For the same period in Japan, where the quality of fish is a national priority, 42 incidents involving 4,122 people were recorded. Since 1978, 2 actions by FDA have reduced the frequency of intoxications caused by specific products. A defect action level for histamine in canned tuna resulted in increased industry quality control. Secondly, blocklisting of mahi mahi reduced the level of fish imported to the United States.

7. *Course of Disease and Complications*: The onset of intoxication symptoms is rapid, ranging from immediate to 30 minutes. The duration of the illness is usually 3 hours, but may last several days.

8. *Target Populations*: All humans are susceptible to scombroid poisoning; however, the symptoms can be severe for the elderly and for those taking medications such as isoniazid. Because of the worldwide network for harvesting, processing, and distributing fishery products, the impact of the problem is not limited to specific geographical areas of the United States or consumption pattern. These foods are sold for use in homes, schools, hospitals, and restaurants as fresh, frozen, or processed products.

9. *Food Analysis*: An official method was developed at FDA to determine histamine, using a simple alcoholic extraction and quantitation by fluorescence spectroscopy.

Tetrodotoxin

1. *Name of the Organism*: Tetrodotoxin (anhydrote-trodotoxin 4-epitetrodotoxin, tetrodonic acid)

2. *Nature of Acute Disease*: Pufferfish Poisoning, Tetradon Poisoning, Fugu Poisoning

3. Nature of Disease: Fish poisoning by consumption of members of the order Tetraodontiformes is one of the most violent intoxications from marine species. The gonads, liver, intestines, and skin of pufferfish can contain levels of tetrodotoxin sufficient to produce rapid and violent death. The flesh of many pufferfish may not usually be dangerously toxic. Tetrodotoxin has also been isolated from widely differing animal species, including the California newt, parrotfish, frogs of the genus Atelopus, the blue-ringed octopus, starfish, angelfish, and xanthid crabs. The metabolic source of tetrodotoxin is uncertain. No algal source has been identified, and until recently tetrodotoxin was assumed to be a metabolic product of the host. However, recent reports of the production of tetrodotoxin/anhydrotetrodotoxin by several bacterial species, including strains of the family Vibrionaceae, *Pseudomonas sp.*, and

Photobacterium phosphoreum, point toward a bacterial origin of this family of toxins. These are relatively common marine bacteria that are often associated with marine animals. If confirmed, these findings may have some significance in toxicoses that have been more directly related to these bacterial species.

4. *Diagnosis of Human Illness*: The diagnosis of pufferfish poisoning is based on the observed symptomology and recent dietary history.

5. *Associated Foods*: Poisonings from tetrodotoxin have been almost exclusively associated with the consumption of pufferfish from waters of the Indo-Pacific ocean regions. Several reported cases of poisonings, including fatalities, involved pufferfish from the Atlantic Ocean, Gulf of Mexico, and Gulf of California. There have been no confirmed cases of poisoning from the Atlantic pufferfish, Spheroides maculatus. However, in one study, extracts from fish of this species were highly toxic in mice. The trumpet shell Charonia sauliae has been implicated in food poisonings, and evidence suggests that it contains a tetrodotoxin derivative. There have been several reported poisonings from mislabelled pufferfish and at least one report of a fatal episode when an individual swallowed a California newt.

6. *Relative Frequency of Disease*: From 1974 through 1983 there were 646 reported cases of pufferfish poisoning in Japan, with 179 fatalities. Estimates as high as 200 cases per year with mortality approaching 50% have been reported. Only a few cases have been reported in the United States, and outbreaks in countries outside the Indo-Pacific area are rare.

7. *Course of Disease and Complications*: The first symptom of intoxication is a slight numbness of the lips and tongue, appearing between 20 minutes to three hours after eating poisonous pufferfish. The next symptom is increasing paraesthesia in the face and extremities, which may be followed by sensations of lightness or floating. Headache, epigastric pain, nausea, diarrhea, and/or vomiting may occur. Occasionally, some reeling or difficulty in walking may occur. The second stage of the intoxication is increasing paralysis. Many victims are unable to move; even sitting may be difficult. There is increasing respiratory distress. Speech is affected, and the victim usually exhibits dyspnea, cyanosis, and hypotension. Paralysis increases and convulsions, mental impairment, and cardiac arrhythmia may occur. The victim, although completely paralyzed, may be conscious and in some cases completely lucid until shortly before death. Death usually occurs within 4 to 6 hours, with a known range of about 20 minutes to 8 hours.

8. *Target Populations*: All humans are susceptible to tetrodotoxin poisoning. This toxicosis may be avoided by not consuming pufferfish or other animal species containing tetrodotoxin. Most other animal species known to contain

tetrodotoxin are not usually consumed by humans. Poisoning from tetrodotoxin is of major public health concern primarily in Japan, where "fugu" is a traditional delicacy. It is prepared and sold in special restaurants where trained and licensed individuals carefully remove the viscera to reduce the danger of poisoning. Importation of pufferfish into the United States is not generally permitted, although special exceptions may be granted. There is potential for misidentification and/or mislabelling, particularly of prepared, frozen fish products.

9. *Food Analysis*: The mouse bioassay developed for paralytic shellfish poisoning (PSP) can be used to monitor tetrodotoxin in pufferfish and is the current method of choice. An HPLC method with post- column reaction with alkali and fluorescence has been developed to determine tetrodotoxin and its associated toxins. The alkali degradation products can be confirmed as their trimethylsilyl derivatives by gas chromatography/mass spectrometry. These chromatographic methods have not yet been validated.

Mushroom Toxins

1. *Name of the Organism*: Amanitin, Gyromitrin, Orellanine, Muscarine, Ibotenic Acid, Muscimol, Psilocybin, Coprine

2. *Nature of Acute Disease*: Mushroom Poisoning, Toadstool Poisoning Protoplasmic Neurotoxins Gastrointinstinal Irritants Disulfiram-like Miscellaneous Types of Poisons.

– Mushroom poisoning is caused by the consumption of raw or cooked fruiting bodies (mushrooms, toadstools) of a number of species of higher fungi. The term toadstool (from the German Todesstuhl, death's stool) is commonly given to poisonous mushrooms, but for individuals who are not experts in mushroom identification there are generally no easily recognizable differences between poisonous and nonpoisonous species. Old wives' tales notwithstanding, there is no general rule of thumb for distinguishing edible mushrooms and poisonous toadstools. The toxins involved in mushroom poisoning are produced naturally by the fungi themselves, and each individual specimen of a toxic species should be considered equally poisonous. Most mushrooms that cause human poisoning cannot be made nontoxic by cooking, canning, freezing, or any other means of processing. Thus, the only way to avoid poisoning is to avoid consumption of the toxic species. Poisonings in the United States occur most commonly when hunters of wild mushrooms (especially novices) misidentify and consume a toxic species, when recent immigrants collect and consume a poisonous American species that closely resembles an edible wild mushroom from their native land, or when mushrooms that contain psychoactive compounds are intentionally consumed by persons who desire these effects.

3. *Nature of Disease*: Mushroom poisonings are generally acute and are manifested by a variety of symptoms and prognoses, depending on the amount and species consumed. Because the chemistry of many of the mushroom toxins (especially the less deadly ones) is still unknown and positive identification of the mushrooms is often difficult or impossible, mushroom poisonings are generally categorized by their physiological effects. There are four categories of mushroom toxins: protoplasmic poisons (poisons that result in generalized destruction of cells, followed by organ failure); neurotoxins (compounds that cause neurological symptoms such as profuse sweating, coma, convulsions, hallucinations, excitement, depression, spastic colon); gastrointestinal irritants (compounds that produce rapid, transient nausea, vomiting, abdominal cramping, and diarrhea); and disulfiram-like toxins. Mushrooms in this last category are generally nontoxic and produce no symptoms unless alcohol is consumed within 72 hours after eating them, in which case a short-lived acute toxic syndrome is produced.

4. *Diagnosis of Human Illness*: A clinical testing procedure is currently available only for the most serious types of mushroom toxins, the amanitins. The commercially available method uses a 3H- radioimmunoassay (RIA) test kit and can detect sub-nanogram levels of toxin in urine and plasma. Unfortunately, it requires a 2- hour incubation period, and this is an excruciating delay in a type of poisoning which the clinician generally does not see until a day or two has passed. A 125I-based kit which overcomes this problem has recently been reported, but has not yet reached the clinic. A sensitive and rapid HPLC technique has been reported in the literature even more recently, but it has not yet seen clinical application. Since most clinical laboratories in this country do not use even the older RIA technique, diagnosis is based entirely on symptomology and recent dietary history. Despite the fact that cases of mushroom poisoning may be broken down into a relatively small number of categories based on symptomatology, positive botanical identification of the mushroom species consumed remains the only means of unequivocally determining the particular type of intoxication involved, and it is still vitally important to obtain such accurate identification as quickly as possible. Cases involving ingestion of more than one toxic species in which one set of symptoms masks or mimics another set are among many reasons for needing this information. Unfortunately, a number of factors (not discussed here) often make identification of the causative mushroom impossible. In such cases, diagnosis must be based on symptoms alone. In order to rule out other types of food poisoning and to conclude

that the mushrooms eaten were the cause of the poisoning, it must be established that everyone who ate the suspect mushrooms became ill and that no one who did not eat the mushrooms became ill. Wild mushrooms eaten raw, cooked, or processed should always be regarded as prime suspects. After ruling out other sources of food poisoning and positively implicating mushrooms as the cause of the illness, diagnosis may proceed in two steps. As described above, the protoplasmic poisons are the most likely to be fatal or to cause irreversible organ damage. In the case of poisoning by the deadly Amanitas, important laboratory indicators of liver (elevated LDH, SGOT, and bilirubin levels) and kidney (elevated uric acid, creatinine, and BUN levels) damage will be present. Unfortunately, in the absence of dietary history, these signs could be mistaken for symptoms of liver or kidney impairment as the result of other causes (e.g., viral hepatitis). It is important that this distinction be made as quickly as possible, because the delayed onset of symptoms will generally mean that the organ has already been damaged. The importance of rapid diagnosis is obvious: victims who are hospitalized and given aggressive support therapy almost immediately after ingestion have a mortality rate of only 10%, whereas those admitted 60 or more hours after ingestion have a 50-90% mortality rate. A recent report indicates that amanitins are observable in urine well before the onset of any symptoms, but that laboratory tests for liver dysfunction do not appear until well after the organ has been damaged.

5. *Associated Foods*: Mushroom poisonings are almost always caused by ingestion of wild mushrooms that have been collected by nonspecialists (although specialists have also been poisoned). Most cases occur when toxic species are confused with edible species, and a useful question to ask of the victims or their mushroom-picking benefactors is the identity of the mushroom they thought they were picking. In the absence of a well- preserved specimen, the answer to this question could narrow the possible suspects considerably.

Intoxication has also occurred when reliance was placed on some folk method of distinguishing poisonous and safe species. Outbreaks have occurred after ingestion of fresh, raw mushrooms, stir-fried mushrooms, home-canned mushrooms, mushrooms cooked in tomato sauce (which rendered the sauce itself toxic, even when no mushrooms were consumed), and mushrooms that were blanched and frozen at home. Cases of poisoning by home-canned and frozen mushrooms are especially insidious because a single outbreak may easily become a multiple outbreak when the preserved toadstools are carried to another location and consumed at another time. Specific cases of mistaken mushroom identity appears frequently. The Early False Morel *Gyromitra esculenta* is easily confused with the true Morel *Morchella esculenta*, and poisonings have occurred after consumption of fresh or cooked *Gyromitra*. *Gyromitra* poisonings have also occurred after ingestion of commercially available "morels" contaminated with

G. esculenta. The commercial sources for these fungi (which have not yet been successfully cultivated on a large scale) are field collection of wild morels by semiprofessionals. Cultivated commercial mushrooms of whatever species are almost never implicated in poisoning outbreaks unless there are associated problems such as improper canning (which lead to bacterial food poisoning). Producers of mild gastroenteritis are too numerous to list here, but include members of many of the most abundant genera, including *Agaricus, Boletus, Lactarius, Russula, Tricholoma, Coprinus, Pluteus,* and others. The Inky Cap Mushroom (*Coprinus atrimentarius*) is considered both edible and delicious, and only the unwary who consume alcohol after eating this mushroom need be concerned. Some other members of the genus *Coprinus* (Shaggy Mane, *C. comatus;* Glistening Inky Cap,

C. micaceus, and others) and some of the larger members of the *Lepiota* family such as the Parasol Mushroom (*Leucocoprinus procera*) do not contain coprine and do not cause this effect. The potentially deadly Sorrel Webcap Mushroom (*Cortinarius orellanus*) is not easily distinguished from nonpoisonous webcaps belonging to the same distinctive genus, and all should be avoided.Most of the psychotropic mushrooms (*Inocybe* spp., *Conocybe* spp., *Paneolus* spp., *Pluteus* spp.) are in general appearance small, brown, and leathery (the so-called "Little Brown Mushrooms" or LBMs) and relatively unattractive from a culinary standpoint. The Sweat Mushroom (*Clitocybe dealbata*) and the Smoothcap Mushroom (*Psilocybe cubensis*) are small, white, and leathery. These small, unattractive mushrooms are distinctive, fairly unappetizing, and not easily confused with the fleshier fungi normally considered edible. Intoxications associated with them are less likely to be accidental, although both *C. dealbata* and *Paneolus foenisicii* have been found growing in the same fairy ring area as the edible (and

choice) Fairy Ring Mushroom (*Marasmius oreades*) and the Honey Mushroom (*Armillariella mellea*), and have been consumed when the picker has not carefully examined every mushroom picked from the ring. Psychotropic mushrooms, which are larger and therefore more easily confused with edible mushrooms, include the Showy Flamecap or Big Laughing Mushroom (*Gymnopilus spectabilis*), which has been mistaken for Chanterelles (*Cantharellus* spp.) and for *Gymnopilus ventricosus* found growing on wood of conifers in western North America. The Fly Agaric (*Amanita muscaria*) and Panthercap (*Amanita pantherina*) mushrooms are large, fleshy, and colorful. Yellowish cap colors on some varieties of the Fly Agaric and the Panthercap are similar to the edible Caesar's Mushroom (*Amanita caesarea*), which is considered a delicacy in Italy. Another edible yellow capped mushroom occasionally confused with yellow *A. muscaria* and *A. pantherina* varieties are the Yellow Blusher (*Amanita flavorubens*). Orange to yellow-orange *A. muscaria* and *A. pantherina* may also be confused with the Blusher (*Amanita rubescens*) and the Honey Mushroom (*Armillariella mellea*). White to pale forms of *A. muscaria* may be confused with edible field mushrooms (*Agaricus* spp.). Young (button stage) specimens of *A. muscaria* have also been confused with puffballs.

6. *Relative Frequency of Disease*: Accurate figures on the relative frequency of mushroom poisonings are difficult to obtain. For the 5-year period between 1976 and 1981, 16 outbreaks involving 44 cases were reported to the Centers for Disease Control in Atlanta. The number of unreported cases is, of course, unknown. Cases are sporadic and large outbreaks are rare. Poisonings tend to be grouped in the spring and fall when most mushroom species are at the height of their fruiting stage. While the actual incidence appears to be very low, the potential exists for grave problems. Poisonous mushrooms are not limited in distribution as are other poisonous organisms (such as dinoflagellates). Intoxications may occur at any time and place, with dangerous species occurring in habitats ranging from urban lawns to deep woods. As Americans become more adventurous in their mushroom collection and consumption, poisonings are likely to increase.

7. *Course of Disease and Complications*: The normal course of the disease varies with the dose and the mushroom species eaten. Each poisonous species contains one or more toxic compounds which are unique to few other species. Therefore, cases of mushroom poisonings generally do not resembles each other unless they are caused by the same or very closely related mushroom species. Almost all mushroom poisonings may be grouped in one of the categories outlined above.

Protoplasmic Poisons

– *Amatoxins*: Several mushroom species, including the Death Cap or Destroying Angel (*Amanita phalloides, A. virosa*), the Fool's Mushroom (*A. verna*) and several of their relatives, along with the Autumn Skullcap (*Galerina autumnalis*) and some of its relatives, produce a family of cyclic octapeptides called amanitins. Poisoning by the amanitins is characterized by a long latent period (range 6-48 hours, average 6-15 hours) during which the patient shows no symptoms. Symptoms appear at the end of the latent period in the form of sudden, severe seizures of abdominal pain, persistent vomiting and watery diarrhea, extreme thirst, and lack of urine production. If this early phase is survived, the patient may appear to recover for a short time, but this period will generally be followed by a rapid and severe loss of strength, prostration, and pain-caused restlessness. Death in 50-90% of the cases from progressive and irreversible liver, kidney, cardiac, and skeletal muscle damage may follow within 48 hours (large dose), but the disease more typically lasts 6 to 8 days in adults and 4 to 6 days in children. Two or three days after the onset of the later phase, jaundice, cyanosis, and coldness of the skin occur. Death usually follows a period of coma and occasionally convulsions. If recovery occurs, it generally requires at least a month and is accompanied by enlargement of the liver. Autopsy will usually reveal fatty degeneration and necrosis of the liver and kidney.

– *Hydrazines*: Certain species of False Morel (*Gyromitra esculenta* and *G. gigas*) contain the protoplasmic poison gyromitrin, a volatile hydrazine derivative. Poisoning by this toxin superficially resembles *Amanita* poisoning but is less severe. There is generally a latent period of 6 - 10 hours after ingestion during which no symptoms are evident, followed by sudden onset of abdominal discomfort (a feeling of fullness), severe headache, vomiting, and sometimes diarrhea. The toxin affects primarily the liver, but there are additional disturbances to blood cells and the central nervous system. The mortality rate is relatively low (2-4%). Poisonings with symptoms almost identical to those produced by *Gyromitra* have also been reported after ingestion of the Early False Morel (*Verpa bohemica*). The toxin is presumed to be related to gyromitrin but has not yet been identified.

– *Orellanine*: The final type of protoplasmic poisoning is caused by the Sorrel Webcap mushroom (*Cortinarius orellanus*) and some of its relatives. This mushroom produces orellanine, which causes a type of poisoning characterized by an extremely long asymptomatic latent period of 3 to 14 days. An intense, burning thirst (polydipsia) and excessive urination (polyuria) are the first symptoms. This may be followed by nausea, headache, muscular pains, chills, spasms, and loss of consciousness. In severe cases, severe renal tubular necrosis and kidney failure may result in death (15%) several weeks after the poisoning. Fatty degeneration of the liver and severe inflammatory changes in the intestine accompany the renal damage, and recovery in less severe cases may require several months. NEUROTOXINS Poisonings by mushrooms that cause neurological problems may be divided into three groups, based on the type of symptoms produced, and named for the substances responsible for these symptoms.

– *Muscarine Poisoning*: Ingestion of any number of *Inocybe* or *Clitocybe* species (e.g., *Inocybe geophylla, Clitocybe dealbata*) results in an illness characterized primarily by profuse sweating. This effect is caused by the presence in these mushrooms of high levels (3- 4%) of muscarine. Muscarine poisoning is characterized by increased salivation, perspiration, and lacrimation within 15 to 30 minutes after ingestion of the mushroom. With large doses, these symptoms may be followed by abdominal pain, severe nausea, diarrhea, blurred vision, and labored breathing. Intoxication generally subsides within 2 hours. Deaths are rare, but may result from cardiac or respiratory failure in severe cases.

– *Ibotenic acid/Muscimol Poisoning*: The Fly Agaric (*Amanita muscaria*) and Panthercap (*Amanita pantherina*) mushrooms both produce ibotenic acid and muscimol. Both substances produce the same effects, but muscimol is approximately 5 times more potent than ibotenic acid. Symptoms of poisoning generally occur within 1 - 2 hours after ingestion of the mushrooms. An initial abdominal discomfort may be present or absent, but the chief symptoms are drowsiness and dizziness (sometimes accompanied by sleep), followed by a period of hyperactivity, excitability, illusions, and delirium. Periods of drowsiness may alternate with periods of excitement, but symptoms generally fade within a few hours. Fatalities rarely occur in adults, but in children, accidental consumption of large quantities of these mushrooms may cause convulsions, coma, and other neurologic problems for up to 12 hours.

–*Psilocybin Poisoning*: A number of mushrooms belonging to the genera *Psilocybe, Panaeolus, Copelandia, Gymnopilus, Conocybe,* and *Pluteus,* when ingested, produce a syndrome similar to alcohol intoxication (sometimes accompanied by hallucinations). Several of these mushrooms (e.g., *Psilocybe cubensis, P. mexicana, Conocybe cyanopus*) are eaten for their psychotropic effects in religious ceremonies of certain native American tribes, a practice which dates to the pre- Columbian era. The toxic effects are caused by psilocin and psilocybin. Onset of symptoms is usually rapid and the effects generally subside within 2 hours. Poisonings by these mushrooms are rarely fatal in adults and may be distinguished from ibotenic acid poisoning by the absence of drowsiness or coma. The most severe cases of psilocybin poisoning occur in small children, where large doses may cause the hallucinations accompanied by fever, convulsions, coma, and death. These mushrooms are generally small, brown, nondescript, and not particularly fleshy; they are seldom mistaken for food fungi by innocent hunters of wild mushrooms. Poisonings caused by intentional ingestion of these mushrooms by people with no legitimate religious justification must be handled with care, since the only cases likely to be seen by the physician are overdoses or intoxications caused by a combination of the mushroom and some added psychotropic substance (such as PCP).

– *Gastrointestinal Irritants*: Numerous mushrooms, including the Green Gill (*Chlorophyllum molybdites*), Gray Pinkgill (*Entoloma lividum*), Tigertop

(*Tricholoma pardinum*), Jack O'Lantern (*Omphalotus illudens*), Naked Brimcap (*Paxillus involutus*), Sickener (*Russula emetica*), Early False Morel (*Verpa bohemica*), Horse mushroom (*Agaricus arvensis*) and Pepper bolete (*Boletus piperatus*), contain toxins that can cause gastrointestinal distress, including but not limited to nausea, vomiting, diarrhea, and abdominal cramps. In many ways these symptoms are similar to those caused by the deadly protoplasmic poisons. The chief and diagnostic difference is that poisonings caused by these mushrooms have a rapid onset, rather than the delayed onset seen in protoplasmic poisonings. Some mushrooms (including the first five species mentioned above) may cause vomiting and/ or diarrhea which lasts for several days. Fatalities caused by these mushrooms are relatively rare and are associated with dehydration and electrolyte imbalances caused by diarrhea and vomiting, especially in debilitated, very young, or very old patients. Replacement of fluids and other appropriate supportive therapy will prevent death in these cases. The chemistry of the toxins responsible for this type of poisoning is virtually unknown, but may be related to the presence in some mushrooms of unusual sugars, amino acids, peptides, resins, and other compounds.

– *Disulfiram-Like Poisoning*: The Inky Cap Mushroom (*Coprinus atramentarius*) is most commonly responsible for this poisoning, although a few other species have also been implicated. A complicating factor in this type of intoxication is that this species is generally considered edible (i.e., no illness results when eaten in the absence of alcoholic beverages). The mushroom produces an unusual amino acid, coprine, which is converted to cyclopropanone hydrate in the human body. This compound interferes with the breakdown of alcohol, and consumption of alcoholic beverages within 72 hours after eating it will cause headache, nausea and vomiting, flushing, and cardiovascular disturbances that last for 2 - 3 hours.

– *Miscellaneous Poisonings*: Young fruiting bodies of the sulfur shelf fungus *Laetiporus sulphureus* are considered edible. However, ingestion of this shelf fungus has caused digestive upset and other symptoms in adults and visual hallucinations and ataxia in a child.

8. *Target Populations*: All humans are susceptible to mushroom toxins. The poisonous species are ubiquitous, and geographical restrictions on types of poisoning that may occur in one location do not exist (except for some of the hallucinogenic LBMs, which occur primarily in the American southwest and southeast). Individual specimens of poisonous mushrooms are also characterized by individual variations in toxin content based on genetics, geographic location, and growing conditions. Intoxications may thus be more or less serious, depending not on the number of mushrooms consumed, but on the dose of toxin delivered. In addition, although most cases of

poisoning by higher plants occur in children, toxic mushrooms are consumed most often by adults. Occasional accidental mushroom poisonings of children and pets have been reported, but adults are more likely to actively search for and consume wild mushrooms for culinary purposes. Children are more seriously affected by the normally nonlethal toxins than are adults and are more likely to suffer very serious consequences from ingestion of relatively smaller doses. Adults who consume mushrooms are also more likely to recall what was eaten and when, and are able to describe their symptoms more accurately than are children. Very old, very young, and debilitated persons of both sexes are more likely to become seriously ill from all types of mushroom poisoning, even those types which are generally considered to be mild. Many idiosyncratic adverse reactions to mushrooms have been reported. Some mushrooms cause certain people to become violently ill, while not affecting others who consumed part of the same mushroom cap. Factors such as age, sex, and general health of the consumer do not seem to be reliable predictors of these reactions, and they have been attributed to allergic or hypersensitivity reactions and to inherited inability of the unfortunate victim to metabolize certain unusual fungal constituents (such as the uncommon sugar, trehalose). These reactions are probably not true poisonings as the general population does not seem to be affected.

9. *Food Analysis*: The mushroom toxins can with difficulty be recovered from poisonous fungi, cooking water, stomach contents, serum, and urine. Procedures for extraction and quantitation are generally elaborate and time-consuming, and the patient will in most cases have recovered by the time an analysis is made on the basis of toxin chemistry. The exact chemical natures of most of the toxins that produce milder symptoms are unknown. Chromatographic techniques (TLC, GLC, HPLC) exist for the amanitins, orellanine, muscimol/ibotenic acid, psilocybin, muscarine, and the gyromitrins. The amanitins may also be determined by commercially available 3H-RIA kits. The most reliable means of diagnosing a mushroom poisoning remains botanical identification of the fungus that was eaten. An accurate pre-ingestion determination of species will also prevent accidental poisoning in 100% of cases. Accurate post- ingestion analyses for specific toxins when no botanical identification is possible may be essential only in cases of suspected poisoning by the deadly *Amanitas*, since prompt and aggressive therapy (including lavage, activated charcoal, and plasmapheresis) can greatly reduce the mortality rate

Aflatoxins

1. *Name of the Organism*: Aflatoxins

2. *Nature of Acute Disease*: Aflatoxicosis Aflatoxicosis is poisoning that results from ingestion of aflatoxins in contaminated food or feed. The aflatoxins are a group of structurally related toxic compounds produced by certain strains of the fungi *Aspergillus flavus* and *A. parasiticus*. Under favorable conditions of temperature and humidity, these fungi grow on certain foods and feeds, resulting in the production of aflatoxins. The most pronounced contamination has been encountered in tree nuts, peanuts, and other oilseeds, including corn and cottonseed. The major aflatoxins of concern are designated B1, B2, G1, and G2. These toxins are usually found together in various foods and feeds in various proportions; however, aflatoxin B1 is usually predominant and is the most toxic. When a commodity is analyzed by thin-layer chromatography, the aflatoxins separate into the individual components in the order given above; however, the first two fluoresce blue when viewed under ultraviolet light and the second two fluoresce green. Aflatoxin M a major metabolic product of aflatoxin B1 in animals and is usually excreted in the milk and urine of dairy cattle and other mammalian species that have consumed aflatoxin-contaminated food or feed.

3. *Nature of Disease*: Aflatoxins produce acute necrosis, cirrhosis, and carcinoma of the liver in a number of animal species; no animal species is resistant to the acute toxic effects of aflatoxins; hence it is logical to assume that humans may be similarly affected. A wide variation in LD50 values has been obtained in animal species tested with single doses of aflatoxins. For most species, the LD50 value ranges from 0.5 to 10 mg/kg body weight. Animal species respond differently in their susceptibility to the chronic and acute toxicity of aflatoxins. The toxicity can be influenced by environmental factors, exposure level, and duration of exposure, age, health, and nutritional status of diet. Aflatoxin B1 is a very potent carcinogen in many species, including nonhuman primates, birds, fish, and rodents. In each species, the liver is the primary target organ of acute injury. Metabolism plays a major role in determining the toxicity of aflatoxin B1; studies show that this aflatoxion requires metabolic activation to exert its carcinogenic effect, and these effects can be modified by induction or inhibition of the mixed function oxidase system.

4. *Diagnosis of Human Illness*: Aflatoxicosis in humans has rarely been reported; however, such cases are not always recognized. Aflatoxicosis may be suspected when a disease outbreak exhibits the following characteristics: · the cause is not readily identifiable · the condition is not transmissible · s y n d r o m e s may be associated with certain batches of food · treatment with antibiotics or other drugs has little effect · the outbreak may be seasonal, i.e., weather conditions may affect mold growth. The adverse effects of aflatoxins

in animals (and presumably in humans) have been categorized in two general forms.A. (Primary) Acute aflatoxicosis is produced when moderate to high levels of aflatoxins are consumed. Specific, acute episodes of disease ensue may include hemorrhage, acute liver damage, edema, alteration in digestion, absorption and/or metabolism of nutrients, and possibly death.B. (Primary) Chronic aflatoxicosis results from ingestion of low to moderate levels of aflatoxins. The effects are usually subclinical and difficult to recognize. Some of the common symptoms are impaired food conversion and slower rates of growth with or without the production of an overt aflatoxin syndrome.

5. *Associated Foods*: In the United States, aflatoxins have been identified in corn and corn products, peanuts and peanut products, cottonseed, milk, and tree nuts such as Brazil nuts, pecans, pistachio nuts, and walnuts. Other grains and nuts are susceptible but less prone to contamination.

6. *Relative Frequency of Disease*: The relative frequency of aflatoxicosis in humans in the United States is not known. No outbreaks have been reported in humans. Sporadic cases have been reported in animals.

7. *Course of Disease and Complications*: In well-developed countries, aflatoxin contamination rarely occurs in foods at levels that cause acute aflatoxicosis in humans. In view of this, studies on human toxicity from ingestion of aflatoxins have focused on their carcinogenic potential. The relative susceptibility of humans to aflatoxins is not known, even though epidemiological studies in Africa and Southeast Asia, where there is a high incidence of hepatoma, have revealed an association between cancer incidence and the aflatoxin content of the diet. These studies have not proved a cause-effect relationship, but the evidence suggests an association. One of the most important accounts of aflatoxicosis in humans occurred in more than 150 villages in adjacent districts of two neighboring states in northwest India in the fall of 1974. According to one report of this outbreak, 397 persons were affected and 108 persons died.

In this outbreak, contaminated corn was the major dietary constituent, and aflatoxin levels of 0.25 to 15 mg/kg were found. The daily aflatoxin B1 intake was estimated to have been at least 55 ug/kg body weight for an undetermined number of days. The patients experienced high fever, rapid progressive jaundice, edema of the limbs, pain, vomiting, and swollen livers. One investigator reported a peculiar and very notable feature of the outbreak: the appearance of signs of disease in one village population was preceded by a similar disease in domestic dogs, which was usually fatal.

Histopathological examination of humans showed extensive bile duct proliferation and periportal fibrosis of the liver together with gastrointestinal hemorrhages. A 10-year follow-up of the Indian outbreak found the survivors fully recovered with no ill effects from the experience.A second outbreak of aflatoxicosis was reported from Kenya in 1982. There were 20 hospital admissions with a 60% mortality; daily aflatoxin intake was estimated to be at least 38 ug/ kg body weight for an undetermined number of days.In a deliberate suicide attempt, a laboratory worker ingested 12 ug/kg body weight of aflatoxin B1 per day over a 2-day period and 6 months later, 11 ug/kg body weight per day over a 14-day period. Except for transient rash, nausea and headache, there were no ill effects; hence, these levels may serve as possible no-effect levels for aflatoxin B1 in humans. In a 14-year follow-up, a physical examination and blood chemistry, including tests for liver function, were normal.

8. *Target Populations*: Although humans and animals are susceptible to the effects of acute aflatoxicosis, the chances of human exposure to acute levels of aflatoxin is remote in well-developed countries. In undeveloped countries, human susceptibility can vary with age, health, and level and duration of exposure.

9. *Food Analysis*: Many chemical procedures have been developed to identify and measure aflatoxins in various commodities. The basic steps include extraction, lipid removal, cleanup, separation and quantification. Depending on the nature of the commodity, methods can sometimes be simplified by omitting unnecessary steps. Chemical methods have been developed for peanuts, corn, cottonseed, various tree nuts, and animal feeds. Chemical methods for aflatoxin in milk and dairy products are far more sensitive than for the above commodities because the aflatoxin M animal metabolite is usually found at much lower levels (ppb and ppt).

Phytohaemagglutinin

1. *Name of the Organism*: Phytohaemagglutinin (Kidney Bean Lectin) This compound, a lectin or hemagglutinin, has been used by immunologists for years to trigger DNA synthesis in T lymphocytes, and more recently, to activate latent human immunodeficiency virus type 1 (HIV-1, AIDS virus) from human peripheral lymphocytes. Besides inducing mitosis, lectins are known for their ability to agglutinate many mammalian red blood cell types, alter cell membrane transport systems, alter cell permeability to proteins, and generally interfere with cellular metabolism.

2. *Nature of Acute Disease*: Red Kidney Bean (*Phaseolus vulgaris*) Poisoning, Kinkoti Bean Poisoning, and possibly other names.

3. Nature of Disease: The onset time from consumption of raw or undercooked kidney beans to symptoms varies from between 1 to 3 hours.

Onset is usually marked by extreme nausea, followed by vomiting, which may be very severe. Diarrhea develops somewhat later (from one to a few hours), and some persons report abdominal pain. Some persons have been hospitalized, but recovery is usually rapid (3 - 4 h after onset of symptoms) and spontaneous.

4. *Diagnosis of Human Illness*: Diagnosis is made on the basis of symptoms, food history, and the exclusion of other rapid onset food poisoning agents (e.g., *Bacillus cereus, Staphylococcus aureus*, arsenic, mercury, lead, and cyanide).

5. *Associated Foods*: Phytohaemagglutinin, the presumed toxic agent, is found in many species of beans, but it is in highest concentration in red kidney beans (*Phaseolus vulgaris*). The unit of toxin measure is the hemagglutinating unit (hau). Raw kidney beans contain from 20,000 to 70,000 hau, while fully cooked beans contain from 200 to 400 hau. White kidney beans, another variety of *Phaseolus vulgaris*, contain about one-third the amount of toxin as the red variety; broad beans (*Vicia faba*) contain 5 to 10% the amount that red kidney beans contain. The syndrome is usually caused by the ingestion of raw, soaked kidney beans, either alone or in salads or casseroles. As few as four or five raw beans can trigger symptoms. Several outbreaks have been associated with "slow cookers" or crock pots, or in casseroles which had not reached a high enough internal temperature to destroy the glycoprotein lectin. It has been shown that heating to 80°C may potentiate the toxicity five-fold, so that these beans are more toxic than if eaten raw. In studies of casseroles cooked in slow cookers, internal temperatures often did not exceed 75°C.

6. *Relative Frequency of Disease*: This syndrome has occurred in the United Kingdom with some regularity. Seven outbreaks occurred in the U.K. between 1976 and 1979 and were reviewed (Noah et al. 1980. Br. Med. J. 19 July, 236-7). Two more incidents were reported by Public Health Laboratory Services (PHLS), Colindale, U.K. in the summer of 1988. Reports of this syndrome in the United States are anecdotal and have not been formally published.

7. *Course of Disease and Complications*: The disease course is rapid. All symptoms usually resolve within several hours of onset. Vomiting is usually described as profuse, and the severity of symptoms is directly related to the dose of toxin (number of raw beans ingested). Hospitalization has occasionally resulted, and intravenous fluids may have to be administered. Although of short duration, the symptoms are extremely debilitating.

8. *Target Populations*: All persons, regardless of age or gender, appear to be equally susceptible; the severity is related only to the dose ingested. In the seven outbreaks mentioned above, the attack rate was 100%.

9. Food Analysis: The difficulty in food analysis is that this syndrome is not well known in the medical community. Other possible causes must be eliminated, such as *Bacillus cereus, staphylococcal* food poisoning, or chemical toxicity. If beans are a component of the suspected meal, analysis is quite simple, and based on hemagglutination of red blood cells (hau).

Grayanotoxin

1. Name of the Toxin: Grayanotoxin (formerly known as andromedotoxin, acetylandromedol, and rhodotoxin)

2. Nature of Acute Disease: Honey Intoxication Honey intoxication is caused by the consumption of honey produced from the nectar of rhododendrons. The grayanotoxins cause the intoxication. The specific grayanotoxins vary with the plant species. These compounds are diterpenes, polyhydroxylated cyclic hydrocarbons that do not contain nitrogen. Other names associated with the disease is rhododendron poisoning, mad honey intoxication or grayanotoxin poisoning.

3. Nature of Disease: The intoxication is rarely fatal and generally lasts for no more than 24 hours. Generally the disease induces dizziness, weakness, excessive perspiration, nausea, and vomiting shortly after the toxic honey is ingested. Other symptoms that can occur are low blood pressure or shock, bradyarrhythima (slowness of the heart beat associated with an irregularity in the heart rhythm), sinus bradycardia (a slow sinus rhythm, with a heart rate less than 60), nodal rhythm (pertaining to a node, particularly the atrioventricular node), Wolff- Parkinson-White syndrome (anomalous atrioventricular excitation) and complete atrioventricular block.

4. Diagnosis of Human Illness: The grayanotoxins bind to sodium channels in cell membranes. The binding unit is the group II receptor site, localized on a region of the sodium channel that is involved in the voltage-dependent activation and inactivation. These compounds prevent inactivation; thus, excitable cells (nerve and muscle) are maintained in a state of depolarization, during which entry of calcium into the cells may be facilitated. This action is similar to that exerted by the alkaloids of veratrum and aconite. All of the observed responses of skeletal and heart muscles, nerves, and the central nervous system are related to the membrane effects. Because the intoxication is rarely fatal and recovery generally occurs within 24 hours, intervention may not be required..Severe low blood pressure usually responds to the administration of fluids and correction of bradycardia; therapy with vasopressors (agents that stimulate contraction of the muscular tissue of the capillaries and arteries) is only rarely required. Sinus bradycardia and conduction defects usually respond to atropine therapy; however, in at least one instance the use of a temporary pacemaker was required.

5. *Associated Foods*: In humans, symptoms of poisoning occur after a dose-dependent latent period of a few minutes to two or more hours and include salivation, vomiting, and both circumoral (around or near the mouth) and extremity paresthesia (abnormal sensations). Pronounced low blood pressure and sinus bradycardia develop. In severe intoxication, loss of coordination and progressive muscular weakness result. Extrasystoles (a premature contraction of the heart that is independent of the normal rhythm and arises in response to an impulse in some part of the heart other than the sinoatrial node; called also premature beat) and ventricular tachycardia (an abnormally rapid ventricular rhythm with aberrant ventricular excitation, usually in excess of 150 per minute) with both atrioventricular and intraventricular conduction disturbances also may occur. Convulsions are reported occasionally.

6. *Relative Frequency of Disease*: Grayanotoxin poisoning most commonly results from the ingestion of grayanotoxin- contaminated honey, although it may result from the ingestion of the leaves, flowers, and nectar of rhododendrons. Not all rhododendrons produce grayanotoxins. *Rhododendron ponticum* grows extensively on the mountains of the eastern Black Sea area of Turkey. This species has been associated with honey poisoning since 401 BC. A number of toxin species are native to the United States. Of particular importance are the western azalea (*Rhododendron occidentale*) found from Oregon to southern California, the California rosebay (*Rhododendron macrophyllum*) found from British Columbia to central California, and *Rhododendron albiflorum* found from British Columbia to Oregon and in Colorado. In the eastern half of the United States grayanotoxin-contaminated honey may be derived from other members of the botanical family Ericaceae, to which rhododendrons belong. Mountain laurel (*Kalmia latifolia*) and sheep laurel (*Kalmia angustifolia*) are probably the most important sources of the toxin.

7. *Course of Disease and Complications*: Grayanotoxin poisoning in humans is rare. However, cases of honey intoxication should be anticipated everywhere. Some may be ascribed to a increase consumption of imported honey. Others may result from the ingestion of unprocessed honey with the increased desire of natural foods in the American diet.

8. *Target Populations*: All people are believed to be susceptible to honey intoxication. The increased desire of the American public for natural (unprocessed) foods, may result in more cases of grayanotoxin poisoning. Individuals who obtain honey from farmers who may have only a few hives

are at increased risk. The pooling of massive quantities of honey during commercial processing generally dilutes any toxic substance.

9. *Food Analysis*: The grayanotoxins can be isolated from the suspect commodity by typical extraction procedures for naturally occurring terpenes. The toxins are identified by thin layer chromatography.

Mold Growth and Mycotoxin Production Molds of Interest

- Field fungi - require 20-25% moisture
  - Alternaria
  - Fusarium
  - Cladosporium
  - Helminthosporium
- Storage fungi - require 13-18% moisture
  - Aspergillus
  - Penicillium
- Advanced decay fungi - require 20-25% moisture
  - Fusarium Chaetomium
- Noted for mycotoxin production

Growth Conditions

- aw
  - Grow at lower aw than most bacteria, yeasts
  - Minimum depends on temperature, other factors
  - Minimum ~0.80-0.90 for most general %
- Temperature
  - Most grow at temps <25C
  - Some grow close to 0C
  - Interaction with aw, nutrients
- Substrates

– Can utilize complex carbohydrates

– Don't require rich medium

– Better growth if they can penetrate host

Mycotoxin Production

- Relationship to growth
  - Usually need growth before toxin produced
  - Secondary metabolites: accumulate after growth

- Moisture content
- Certain level required for production
- Level varies with commodity and toxin
- →keep cereal grains below 16% moisture
- →keep nuts and cottonseed below 5-8%
- Temperature
  - Aspergillus usually requires >10C for toxin prod
  - Other molds produce toxin to 0C
  - Cycling of temp. can affect production (+ or -)
  - →store below 5C to prevent aflatoxin prod.
  - →store at or below 0C to prevent patulin, zearalenone
- Substrates (commodities)
  - Plant materials favor toxin production
    a. Various nuts
    b. Cottonseed
    c. Corn
    d. Sorghum
    e. Millet
    f. Grains
  - Herbs and spices resist molds and toxin prod
  - Consider most foods to be possible substrates
  - Aflatoxin prod. needs zinc (unavail. in raw soybeans)

Controlling Growth and Toxin Production

- Damage or stress increases contamination
  - Drought → irrigation, drought-resistant cultivars
  - Insects → pesticides, resistant cultivars
  - Mechanical damage
- Competition restricts growth and toxin production
  - Other invading (nontoxigenic) molds
  - Lactic acid bacteria?
  - Bacteriocin producers?
- Gases in environment affect growth and toxin prod.
  - Aerobic organisms ® require O2
  - Restrict O2 ® depress growth and toxin prod.
  - Don't prevent mycotoxin production until O2 < 1%

- Increased CO_2 decreases toxin prod.
- Need >90% CO_2 to prevent production
- Antifungal agents restrict growth and toxin prod.
  - Sorbates, benzoates, propionate, BHA
  - Herbs, spices, essential oils cinnamon (and oil)
  - Clove (and oil) anise allspice lemon and orange oil

Mycotoxins

Origin

- Synthesis route isoprene to trichodienoids or sesquiterpenes, polyketidesto fusarins, fumonisins, moniliformin, zearalenone Interactions of mycotoxins

History

- Ergotamine etc St. Anthony's Fire, Salem witch trials, French revolution? LSD type alkaloid Claviceps pupurea & C. paspali rye pathogens and other cereal grains
- Turkey X disease in 1960 100,000 birds in England aflatoxins in peanuts from Brazil also in corn, cottonseed, coconut, other nuts B1, G1, M1 metabolite in milk aflatoxin M organisms: Aspergillus flavus, A. parisiticus and some Penicillium sp.; poor competitors with other fungi, survive at low aw (0.83) therefore, see in drought conditions when other fungi are not viable, no usually a pathogen in irrigated crops (popcorn); toxinfor productionaw>0.95 mycotoxin is a complete carcinogen (initiator and promotor); protein(<9% calories) diets suppress aflatoxin carcinogenesis. carcinogen to rats at 15 ppb, trout at 0.8 ppb regulatory levels B1, G1 = 20 ppb, 100 ppb in feedfor dairy cows due to M1 Georgia AF levels over 25 years post-harvest growth and mycotoxin production moisture content for growth in cereals 18-19.5% sampling problems detoxification high pressure ammoniation: never FDA approved but approved for with-in-state commerce in Arizona (where process invented)
- Other Aspergillus mycotoxins Sterigmatocystin carcinogen with Penicillium sp.
- Penicillium mycotoxins Ochratoxin P. verrucosum, also A. ochraceus nephrotoxin in swine, humans? Citrinin also Aspergillus mycotoxin nephrotoxin in swine Patulin Penicillium & Aspergillus species and others food P. expansum in apple and apple products and other neutral (grape, pear) pH foods; reported as high as 45 mg/L; not

found in citrus juices; mycotoxin is heat stable; mycotoxin destroyed by EtOH fermentation to hard apple cider; LD50 = 5-30 mg/kg in mice, rat carcinogen given subcutaneously at 0.2 mg, orally has no effect; US has no regulatory level but 10 countries level set at 50 fg/L; FDA survey in 1993 of juices found >26% exceeding this level; has killed cattle; carcinogen Penicillic acid P. cyclopium death in farm animals?, carcinogen Roquefortine P. roqueforti and others *a*- cyclopiazonic acid P. aurantiogriseum, P. camembertii and Aspergillus

- Fusarium mycotoxins Route of infection silks or anthers versus insect/ mechanical damage Phenotypic identification versus mating populations A-F Fungal contamination versus mycotoxin level moisture content for fungal growth in grain 18.4-33%
- F. graminearum maize, wheat 4-Deoxynivalenol (DON, vomitoxin) vomiting in swine (human?) nivalenol vomiting zearalenone estrogen activity swine breeding problems fusarin C mutagen but not carcinogen
- F. subglutinans maize pathogen moniliformin
- F. moniliforme maize pathogen water-soluble versus organic solvent soluble fumonisin B1, B1, B1, B1 levels in corn, frequency diseases carcinogen Class 2B, equine encephalomalacia >10 ppm, porcine pulmonary edema >25 ppm, esophageal cancer >5 ppm?, poultry feeds >50 ppm food processing meals, flours, grits, beer, pet foods, organic vs others nixtamalization heat-stable nonenzymatic browning fusaric acid fusarin C mutagen
- F. proliferatum maize pathogen water-soluble versus organic solvent soluble fumonisin B1, B1, B1, B1 levels in corn moniliformin
- F. esquiseti wheat, corn fusarochromanone zearalenone estrogen activity
- F. sporotrichioides wheat, corn T-2, HT-2 toxin, T-2 tetraol; T-2 triol dermal necrosis & hemorrhagein cattle, trichothecenes in alimentary toxic aleukia in Soviet Union in WWII, reportedly involved in Yellow Rain in Viet Nam era (Laos), LD50 orally in rats = 4 mg/kg nivalenol vomiting moniliformin, fusarin C
- F. poae wheat, corn T-2, 15-deacetylneosolaniol (NT-20)
- F. culmorum wheat, corn
  culmorin, moniliformin, zearalenone, fusarin C
- F. crookwellense wheat, corn culmorin, zearalenone, *a*- & ß- zearalenol, fusarinC
- F. roseum *a*- & ß-zearalenol estrogenic activity, swine breedingproblems
- Alternaria alternata Alternaria alternata f. sp. lycopersici (AAL) toxin

plant toxicity similar to fumonisins through apoptosis

- Sclerotinia sclerotiorum pink rot disease in celery psoralens skin dermatitis in celery harvesters

SEAFOOD TOXINS

Saxitoxin (Paralytic Shellfish Poisoning)

- Syndrome
  - Sequence of symptoms:
    a. Tingling sensation
    b. Numbness in lips, tongue, legs, arms, neck
    c. Muscular incoordination
    d. Feeling of lightness
    e. Dizziness, weakness, drowsiness, headache
    f. Respiratory distress (diaphragm affected)
    g. Muscular paralysis
    h. Death in 2 to 24 hr
  - Onset: within minutes
  - Duration: normal functions regained in a few days
  - Treatment: no antidote, so treat symptoms artificial respiration
- Occurrence
  - C cause discovered 1927
  - shellfish that have eaten toxic dinoflagellates
    a. Gonyaulax catenella
    b. G. tamarensis
  - Waters off north Pacific and Atlantic coasts of North America, Japan, North Sea
  - Sporadic outbreaks, ~100 cases/year
  - Economic effects as fishing areas avoided
- What happens
  - Unpredictable blooms; some seasonal relation
  - Environmental conditions (temperature, pH, salinity, nutrients, light) favor growth of dinoflagellates
  - "Red tides" - 20,000 cells/ml
  - But 400-500 cells/ml can make shellfish too toxic
  - Toxin concentrated in digestive glands of shellfish

  - Toxic dose builds up rapidly
  - After transfer to fresh water, toxin clears in ~3 weeks
  - No visible difference in toxic shellfish
- Toxin(s)
  - Substituted, dibasic, tetrahydropurine
  - Small molecule (MW = 299)
  - Related toxins have substitutions at key position
  - Very soluble in water
  - Heat-stable: not destroyed by cooking
  - Lethal dose = 1-2 mg
- Activity
  - Neurotoxin
  - Specifically blocks sodium channels in nerve and muscle cell membranes
  - Transfer of nerve impulse stops
- Detection
  - Mouse unit: min. amount of poison that kills a 20-g white mouse in 15 min
  - Standardized mouse assay for routine surveillance C immunoassays with Ab made vs. toxin-protein
  - Constant monitoring of shellfish required

Tetrodotoxin

- Toxigenicity
  - Neurotoxic (syndrome similar to saxitoxin)
  - Prevents conduction of nerve impulses
  - Blocks sodium ion movement into neuron
  - "Plugs" outer end of sodium channel
- Occurrence
  - Wide but species-specific distribution pufferfish salamanders frogs octopus Japanese shellfish
  - Est. >100 cases yearly, ~50% mortality
  - Where does it come from?
    a. Produced by animal itself?
    b. From food chain?
    c. Symbiotic microorganism? ® most likely a symbiotic microbial source because:

d. Cultured pufferfish not toxic
e. Fish became toxic when fed liver, not pure toxin
f. Unusual structure likely from a microorganism

- How can animal tolerate toxin? E evolved sodium channels not sensitive to toxin?

- Toxin
- Unusual structure
- No obvious relation to other natural products
- Lethal dose 1-2 mg

Ciguatera Poisonin

- History
  - Affected early Spanish settlers in Cuba
  - From Spanish "cigua" - name of marine snail
- Syndrome
  - Symptoms: multi-system nausea vomiting diarrhea cramps slow pulse reduced blood pressure headache convulsions tingling numbness hallucinations vertigo
  - Onset: rapid
  - Duration: can persist for months, and reoccur
  - <1% mortality
  - Treatment: symptomatic only
- Occurrence
  - Worldwide, tropical and semitropical areas
  - Sporadic and unpredictable outbreaks
  - Many different fish (grouper, snapper) - unspoiled
  - Estimated 10,000-50,000 cases yearly
  - Impact on in-shore fisheries
- Source of toxin
  - Food chain theory: fish eat toxic dinoflagellate(s)
  - Likely more than one source
  - Secondary toxins may also be present
- Nature of toxin
  - Not well characterized
  - Highly oxygenated lipid
  - ~ 1000 Da

- Not destroyed by cooking or preservation methods (smoking, drying, salting, freezing)
- Detection
  - Bioassay with smooth muscle from guinea pig ileum
  - Mouse lethality
  - ELISA (produce Ab to toxin-protein complex)

Scombroid Poisoning

- Syndrome
  - Symptoms: allergic-like reactions (not all seen) rash localized inflammation nausea vomiting diarrhea hypotension headache tingling flushing itching
  - Onset: several minutes to several hours
  - Duration: few hours (can linger several days) => chemical intoxication
  - Frequently misdiagnosed as allergy
- Epidemiology
  - Worldwide incidence
  - Most common food: scombroid fish (tuna, mackerel)
  - Also other fish (mahi-mahi), cheese
- Cause
  - High levels of histamine ® allergic response
  - Paradox: histamine solution doesn't give symptoms
  - Possible potentiators in fish:
    a. Cadaverine and putrescine (diamines) inhibit diamine oxidase, which detoxifies histamine
    b. Or, they may interfere with protection by mucin
- Source of histamine
  - Decarboxylation of histidine in fish muscle
  - Bacterial sources of histidine decarboxylase
    a. Widely distributed among general
    b. Important in fish: Proteus (Morganella) morganii

Klebsiella pneumoniae Hafnia alvei Clostridium perfringens

- Control: prevent bacterial growth -> low temp. storage

ANATOMY OF MICROORGANISMS AND FOOD BORN INFECTIONS

INTRODUCTION

- The natural microflora of the food
- Microflora contributed by
  - Harvesting
  - Processing
  - Storage
  - Distribution
  - Food preparation

The Numerical Balance of the Components within the Microflora is Determined by

- Properties of the food
- The storage environment of the food
- The properties of the organisms themselves
- The effects of processing

ANATOMY OF THE BACTERIAL CELLS

Bacterial Cells

- Small size, ca. 0.5 to 1.5 mm
- Large surface area in comparison to volume Cell Composition

No "Typical" Cell or Species

- Approximately 70% water
- Dry weight
  - Protein 50%
  - RNA 25%
  - DNA 3%
  - Other 22% (lipids, carbohydrates, metals, etc.)

I. Bacterial Cell Morphology

- Bacilli — rod shaped
- Cocci — spherical
- Vibrio — comma shaped
- Spirilla — spiral shaped

II. STRUCTURES EXTERNAL TO THE CELL WALL

- Flagella
  - Filament
  - Hook
  - Basal structure
- Pili (fimbriae)
  - Common pili
  - Sex (type F) pili
- Glycocalyx
- Axial Filaments

III. Cell Wall (Murein)

Composed of peptidoglycan mucopolysaccharide consisting of a repeatingdisaccharide attached to chains of amino acids (four or five)

- Gram Positive
- Gram Negative

IV. Structures Internal to the Cell Wall

- Cytoplasmic Membrane
- Primarily phospholipids, and proteins
- A phospholipid bilayer, with each layer consisting of a polar (hydrophilic)head consisting of glycerol and phosphate, and a non-polar (hydrophobic)tail of fatty acids
- Mesosomes
- Cytoplasm
- Cytoplasmic Inclusions
- Nuclear Material
  - Nucleoid
  - Plasmids
- Ribosomes

Procaryotic cells are 70S in size, consisting of two subunits a 30Sand a 50S subunit.

INDICATOR MICROORGANISMS

Indicators

Indicators are group of bacteria which indicate potential

contaminationsince we cannot test for every pathogen, look for indicators of contaminationor pathogens.

Indicator Organisms

- A good indicator organism should demonstrate specificity, not be anatural contaminant (e.g., occur predominantly in the intestinal tract)
- Occur in high numbers in the contamination source
- Should be able to survive in a normal environment
- Should be easily and readily detected, even in low numbers

Coliforms

- All aerobic and facultative anaerobic non-sporeforming Gram negativebacilli which ferment lactose to acid and gas within 48 hours at 35°C or37°C
- Capable of growth in the presence of bile salts
- Cytochrome oxidase negative
- Catalase positive
- Beta-galactosidase positive

Enterobacteriaceae

- Non-coliforms:
  - Proteus
  - Providencia
  - Salmonella
  - Serratia
  - Shigella
  - Yersinia
- Coliforms

TOTAL

- Escherichia
- Citrobacter
- Klebsiella
- Enterobacter

Fecal Coliforms

- Escherichia coli

- Citrobacter freundii
- Klebsiella pneumoniae

MUG test

- 96% of *E. coli* strains produce beta-glucoronidase hydrolyzes4- methyl umbelliferyl-beta-D glucuronide to a compound which fluorescesunder long wave UV light
- The majority of *E. coli* strains are MUG positive
- *E. coli* O157:H7 generally does not produce beta-glucoronidase,and is MUG negative

Methods of Detection

Coliforms and Enterobacteriaceae

- Multiple tube fermentation (Most probable number)
  - Lauryl sulfate tryptose broth
  - Brilliant green lactose Bile broth
  - EC broth (thermo-tolerant)
- Direct plating on violet red bile agar, or with added glucose for Enterobacteriaceae
- Membrane filtration
  - m-Endo
  - m-FC

Fecal Streptococci (Enterococci)

- Enterococcus fecalis
- Enterococcus faecium
- Enterococcus gallinarum
- Enterococcus avium

Enterococci

- Either KF broth (MTF) or KF agar (direct plating)
- Azide dextrose broth (MTF)
- Bile esculin azide agar

Other Indicators

- *Clostridium sporogenes* PA3679
- Staphylococcus

  - Staph. aureus
  - Staph. epidermidis
- Clostridium perfringens

BACTERIAL PHYSIOLOGY

I. Metabolism

- Anabolic
- Catabolic

II. Nutritional Patterns

- Energy source
  - Phototroph - light
  - chemotroph - chemical
- Carbon Source
  - Autotrophs - use CO2
  - Hetertrophs - use organic substrates
- Pathways of Energy Production (catabolism)
  - Carbohydrate metabolism provides most of bacterial cells energy
  - Respiration and Fermentation are two main pathways first step in both pathways, for oxidation of glucose to NAD+ is the oxidationof glucose to pyruvic acid; GLYCOLYSIS (Embden-Meyerhof pathway) Respiration pyruvic acid to acetyl Coenzyme
- Acetyl Co A into Krebs cycle Krebs Cycle produces NADH to electron transport chain GTP converted to ADP to ATP Fermentation pyruvic acid to (small amounts of) ATP does not require oxygen or Krebs uses an organic molecule as electron acceptor results in industrially important end products: lactic, acetic, propionicacids ethanol, C02

III. Growth

- Pathways of Energy Production (anabolism)
  - Polysaccharides
  - Amino acids
  - Lipids
- Bacterial cell division
  - Elongation and duplication of DNA
  - Cell wall and plasma membrane begin to form transverse septum, betweenduplicated DNA

- Transverse septum completed
- Cells separate

- Lag phase
- Logarithmic growth
- Stationary
- Death Phase
- Survival Phase
- Mathematical Description of Growth
  - $b = 1 \times 2n$ where b is the bacterial population after n generations, beginning witha single cell
  - $b = B \times 2n$ where B is the initial population
  - $n = 3.3 \log10$ (b/B where n is the number of generations
  - $G = t\ 3.3 \log10 (b/B)$ where G is the generation time and t is the elapsed time between b andB

Sampling Concepts

- What are you sampling for
- Why are you sampling
- How will you use the results

Assessment of Risk Hazard Properties

- Does it contain a sensitive ingredient?
- Will processing destroy hazard?
- Will hazard increase if product is abused?

Assessment of Risk Categories

1. Foods for aged, infants, individuals with predisposing conditions
2. Foods with three hazard properties
3. Foods with two hazard properties
4. Foods with one hazard property
5. Foods with no hazard properties

Assessment of Risk Food and Drug

Administration Categories forsalmonellae

- Category 1. foods that would not normally be subjected to a processlethal to salmonellae, and which are intended for consumption by the aged,infirm, or infants.

- Category 2. foods that would not normally be subjected to a processlethal to salmonellae
- Category 3. foods that would normally be subjected to a process lethalto salmonellae

Sampling Plans

A. Two class plans based on acceptance or rejection of the lot m = critical specification n = number of samples c = number of samplesallowed to exceed the critical specification Example: Product = cooked chicken critical specification: m = coliforms< 10/gram n = 5 per lot c = 2 per lot accept if 3 or more samples meetspecification reject if < 3 samples meet specification

B. Three class plans differentiates between acceptable, margina-llyacceptable, and defective M = second critical specification (defect level)m = critical specification n = number of samples c = number of samplesallowed to exceed the critical specification Example: Product = cooked chicken m = coliforms < 10/gram M = coliforms> 1000/gram n = 5 per lot c = 2 per lot accept if 3 or more samplesmeet specification reject if < 3 samples meet specification accept ifc < 2 reject if c > 2 or if any single sample exceeds M

Sampling Frequency

- Investigational
- Routine
- Reduced

Standards, Guidelines and Specifications

- Standard: a criterion which is part of a law, ordinance or administrativeregulation
- Guideline: a criterion that is often used by the food industry or regulatoryagencies to monitor a process
- Specification: a criterion used as a purchase requirement between abuyer and a vendor

Rationale for Specifications

1. Is there a need for a specification?
2. Will the quality of the raw material make the specification meaningful?
3. Will the effect of further processing and handling increase the hazard?
4. Can process control make a specification unnecessary?
5. What is the cost/benefit ratio of a specification?
6. Is the specification effective?

7. Is the specification attainable by good manufacturing practices?
8. Will a specification encourage the use of objectionable treatments?
9. What action will be taken if a specification is exceeded?
10. What measures are relevant in a specification?
11. Is the methodology of determining the specification sufficiently precise?
12. Which numerical limits should be included in a specification?
13. Which sampling plans should be used?
14. Should there be a mechanism for revision of the specification?

STANDARDS, GUIDELINES, AND SPECIFICATIONS

Grade A Pasteurized Milk Ordinance

Raw Milk

- Standard plate count < 100,000 cfu/ml
- Temperature < 4.4°C * must be cooled to less than 10oC within 2hours of milking

Pasteurization

- 63°C, 30 minutes (long temperature, low time LTLT)
- 72°C, 15 seconds (high temperature, short time HTST)

Pasteurized Milk

- Standard plate count < 20,000 cfu/ml
- Coliforms <10 cfu/ml

Standards, Guidelines and Specifications

- Standard: a criterion which is part of a law, ordinance or administrativeregulation
- Guideline: a criterion that is often used by the food industry or regulatoryagencies to monitor a process
- Specification: a criterion used as a purchase requirement between abuyer and a vendor

Standards, Guidelines and Specifications

- Purposes of specifications:
  - Used to assess the safety, quality, utility or shelf life of a foodor food ingredient

FOOD BORNE DISEASE

Food Borne Disease

- Any disease caused or transmitted by food

Food Borne Intoxication

- Ingestion of a pre-formed toxin

Food Borne Infection

- Ingestion of live organisms which subsequently invade the host

Miscellaneous (Allergies)

- Consumption of a specific compound which triggers an immune response

Conditions Required for Food borne Disease

- Source of bacterium
- Favorable growth environment
- Sufficient time for growth
- Ingestion of sufficient quantity of food to cause illness (Inoculation Dose ID)

Factors Affecting Host Response

- Strain of bacterium involved
- Numbers of bacteria ingested
- Condition of bacterium
- Condition of host

FDA Fecalls

- Class 1 - emergency situation; immediate threat to life, recall allproduct from stores
- Class 2 - priority situation; potential for illness, recall from stores
- Class 3 - routine, no threat to health, mislabeling

Human Digestive Anatomy

1. Oral cavity
2. Esophagus
3. Stomach
4. Small intestine

- Duodenum (25 cm)
- Jejunum (2.5 m)
- Ileum (3.5 m)

5. Large intestine (1.5 m)
   - Cecum
   - Ascending
   - Transverse
   - Descending
   - Sigmoid colon
   - Rectum
   - Anus

Food Born Infections - Salmonella

Salmonella

- Gram negative bacilli, facultative anaerobe, generally motile
- S. typhimurium; enteriditis, typhi, arizonae

Sources

- Intestinal tract of animals
- Polluted water
- Soil

Factors Affecting Growth

- Temperature: optimum 37C
- Generally sensitive to heat, resistant to freezing
- pH: optimum 6.5 - 7.5, minimum 4.0 to 5.5
- Moisture: minimum aw 0.93 to 0.95

Mode of Infection

- Bacteria live and multiply in intestinal tract
- Attach to mucosal lining of intestinal tract and grow
- Bacteria penetrate lining of intestinal wall
- Illness caused by endotoxins which disrupt
- Metabolism of intestinal epithelial cells

Salmonella

- Illness

- Gastroenteritis
  - Onset 6 to 48 hours, typically 24 - 48 hours
  - Symptoms: diarrhea, abdominal cramps, fever, nausea, vomiting, chills
  - Duration: 2 to 3 days
  - Can have acute cases which require hospitalization and/or antibiotictherapy
  - Fatality rate: typically less than 1%
- Enteric Fever
  - Typhoid or paratyphoid fever
  - Onset: 7 to 21 days after infection
  - Symptoms: loss of appetite, headache, high fever
  - Duration: 3 - 6 weeks without treatment
  - Treatment: antibiotics, typically up to 4 weeks;
  - Fatality rate: very high without treatment

Salmonella Cycle

- Animal byproducts (rendering) into feed to animal
- Animals converted to human food and animal byproducts
- Food to humans; humans become infected, both person to food and personto person transfer

Examples

- Host *Serovar*
- Human *S. typhi, S. paratyphi*
- Cattle *S. dublin*
- Swine *S. cholerae-suis*
- Poultry *S. pullorum*
- Sheep *S. abortus-ovis*
- Horse *S. abortus-equi*

Salmonellae Pathogenesis

- Entry into the body (typically oral)
- Bacterial cells which survive the stomach enter the small intestine
  - Invade the intestinal mucosa (microvilli into lamina propria)
  - Engulfed by phagocytes (a portion of the population survives and multiplies)

- Illness caused by degeneration of the villi
- Some cells localize in Peyer's patches (lymphatic tissue)

Anatomy of the Small Intestine

- Mucosa villi
- Sub-mucosa
- Muscularis
- Serosa

Salmonellae Classification

- Approximately 2400 recognized types
- Currently considered as ONE species by the World Health Organization(Salmonella enterica)
- Further subdivided into subspecies
- Subspecies differentiated ito serovars by:
  - Biochemical tests
  - Serology
  - O antigens - the lipopolysaccharide layer
  - K antigens - the flagella
  - Vi antigens - capsule

Salmonella Heat Resistance

- Decimal reduction time at 65oC (D65C)
- Most salmonellaes = 1.5 seconds
- Salmonella seftenberg 775W = 34 seconds D55C
- Most salmonellaes 10 - 12 minutes

Salmonellae Detection

- Pre-enrichment
  - Lactose broth (FDA)
  - Peptone water (USDA)
  - Universal pre-enrichment
- Selective enrichment
  - Selenite cystine broth (FDA, USDA, EEC)
  - Sodium selenite
  - Tetrathionate broth with brilliant green and iodine (USDA, FDA) bilesalts, iodine, brilliant green

  - Rappaport-Vassiliadis R10 (FDA,EEC) malachite green, MgCl2
- Plating
  - Bismuth sulfite (bismuth sulfite, brilliant green)
  - Xylose Lysine Desoxycholate (desoxycholate)
  - Hektoen Enteric (bile salts)

Food Born Infections - Campylobacter

Campylobacter Jejuni

- First recognized as an agent of food borne disease in 1976
- May be the single greatest cause of food borne illness in the U.S.

Sources

- Intestinal tracts of animals and birds
- Polluted water
- Outbreaks associated with: meat, water, raw milk

Factors Affecting Growth

- Micro-aerophillic
- Temperature: survives but does not grow at refrigeration temperatures, range 30 - 45 C

Illness

- Onset: 2- 5 days, but may take up to 10 days
- Duration: 2 - 10 days
- Symptoms: headache, fever, abdominal cramps, profuse watery diarrhea, can have some blood in the stool
- Mechanism: appears to adhere to the intestinal mucosa, intestinal muscosal gel is apparent major coloinization site
- Infective dose: low 100's

Campylobacter Morphology

- Gram negative, non-sporeforming,vibriod (may have 1 - 2 helical turns; may also appear as "S" or "gull-wing" shaped cells
- motile by a cork-screw-like motion

Physiology

- Microaerophilic (5% O2, 10% CO2, 85% N2)

- Carbohydrates are neither fermented or oxidized (energy obtained from amino acids or TCA cycle intermediates)

Species

- C. jejuni
- C. coli
- C. fetus

Related genera

- Arcobacter
- Bdellovibrio
- Helicobacter

Campylobacter Pathogenesis

- Infectious dose: as low as a few hundred
- Two syndromes: motility and chemotaxis (also cell morphology) are involved
- "mild"
  - Profuse watery diarrhea
  - Jejunum and ileum
  - Secretory enterotoxin which disrupts ion transport
- "Severe" or dysenteric
  - Bloody diarrhea
  - Terminal ileum and colon
  - Extensive invasion of mucosa with subsequent necrosis of tissue

Campylobacter Detection

- Enrichment
  - Campylobacter Enrichment Broth
    a. Usually a mixture of antibiotics
    b. Skirrow and Blaser supplements
    c. Vancomycin, polymyxin B, rifampicin
- Plating
  - Similar antibiotics to enrichment
  - Microaerophilic environment
  - Oxygen scavengers (blood, pyruvate, etc.)

Listeria Monocytogenes

- Gram positive, non-spore-forming bacillus
- Generally motile, but motility is temperature dependent, non-motile>30C

Sources

- If not "ubiquitous", then certainly "widely distributedin nature"
- Essentially an environmental isolate

Factors Affecting Growth

- Temperature: optimum 30 - 35C, but will grow very well at refrigerationtemperatures; fairly tolerant to freezing and other food processes
- pH: optimum 6.5 - 7.5, range 4.0 - 9.0

Mode of Infection

- An intracellular pathogen
- Very low infectious dose, ID50 unknown but probably <10 in susceptiblepopulations

Illness

- In most cases, asymptomatic or mild flu-like symptoms
- In immunocompromised, can develop into a variety of symptoms, mostcommonly meningoencephalitis or general septicemia
- Also a cause of spontaneous miscarriage, stillbirth, or developmentof meningitis within a few days of birth
- Fatality rate: can be high if untreated in elderly and children

Species

- L. monocytogenes (+++)
- L. seeligeri (+/-)
- L. welshimeri (+/-)
- L. ivanovii (+/-)
- L. murrayi (-)
- L. innocua (-)

Genera

- Brochothrix

- Carnobacterium
- Kurthia
- Lactobacillus

Listeria Pathogenesis

- Attachment and invasion of intestinal mucosa
- Enters lamina propria
- Engulfed by phagocytes
- Produces listeriolysin O (LLO) to rupture the phagosome and enter thecytoplasm of the phagocyte
- Multiplies in the phagocyte, and invades other tissue from there

Listeria Methodology

- Enrichment in Listeria enrichment broth, University of Vermont (UVM)(naladixic acid)
- Selective enrichment, Fraser broth (USDA) (esculin hydrolysis)
- Selective plating
  - Modified McBride
  - LPM agar
  - Listeria selective medium (Modified Oxford formulation)
  - PALCAM
- Common selective agents: lithium chloride, acriflavin, moxalactam
- *L. monocytogenes* is rhamnose (+), xylose (-)

Listeria Thermal Resistance

- D71.7C 0.9 sec (planktonic) 5.0 sec (intracellular)
- D72C 36 sec (liquid whole egg)

Escherichia coli O157:H7

Escherichia coli O157:H7

- One of several pathogenic *E. coli*'s
- Gram negative bacillus, generally motile
- Enterohemorrhagic (EHEC), also
  - Enteropathogenic (EPEC)
  - Enteroinvasive (EIEC)
  - Enterotoxigenic (ETEC)

Sources

- Epidemiologically associated with ground meat
- Probable reservoir in animals, environment
- Serious outbreaks from improperly treated water

Factors Affecting Growth

- Temperature: rapid growth between 30C and 42C, poor to no growth at44C+
- Survive well during freezing
- pH: optimum near neutrality, but tolerant of a number of conditions,will grow well down to 4.5

Illness

- Onset: 3 - 9 days, average 4 days
- Duration: 2 - 9 days, average 4 days
- Symptoms: abdominal cramps, watery diarrhea, grossly bloody diarrheadescribed as "all blood and no stool", abdominal pain describedas equal in intensity to labor pains
- Hemolytic Uremic Syndrome (HUS)
- hrombotic Thrombocytopenic Purpura (TPP)

Thermal Resistance

Temp. (°F)	Temp. (°C)	D10 (min)
140	60	8.34
145	62.2	2.11
150	65.5	0.53
155	68.3	0.13

Escherichia coli O157:H7

- Colonizes primarily the cecum and colon
- Attaches to villi
- Secretes a heat labile toxin similar to cholera toxin (referred toas SLT, for Shiga-like toxin; encoded on a plasmid)
- Destroys microvilli
- Can progress to hemolytic uremic syndrome in children (caused by toxinsentering blood stream and forming clots)

Escherichia coli O157:H7 Detection

- Pre-enrichment in modified EC broth (bile salts, novobiocin)
- Preliminary screen using ELISA
- Isolate colonies on Sorbitol MacConkey agar
- Confirm suspect colonies
- Biochemical:
  - Escherichia coli Biotype I sorbitol (+), MUG (+)
  - Escherichia coli O157:H7 sorbitol (-), MUG (-)
- Serological:
  - O and H antisera

Shigella

Shigella

- Gram negative bacilli, facultative anaerobe,
- Generally non-motile
- Shigella dysenteriae, flexneri, boydii, sonnei

Sources

- Intestinal tract of animals, especially primates
- polluted water

Factors Affecting Growth

- Temperature: optimum 37°C, (range 10 - 45oC) generally sensitive to heat, resistant to freezing
- pH: optimum 6.5 - 7.5, range 4.0 - 9.0
- Moisture: minimum aw 0.95

Mode of Infection

- Attach to mucosal lining of intestinal tract and grow
- Bacteria penetrate the mucosa
- Bacteria causes extensive tissue damage and
- Intestinal lesions (depending on species)
- Very low infectious dose, ID50 < 10

Food Borne Infections - Shigella

- Illness

- Known as bacillary dysentery
- Onset: usually 7 - 36 hours, depending on inoculum
- Symptoms: (Sh. sonnei) watery diarrhea, possibly fever and chills
- Symptoms: (Sh. dysenteriae) diarrhea with some blood and mucous
- Duration: 5 - 6 days, can last for weeks
- Fatality rate: usually low if treated, can be high if untreated in elderly and children

Shigella - Methods

- Methods generally poor
- re - enrichment followed by selective plating difficult to distinguish from other enetric bacteria

Shigella

Species	Illness
Sh. dysenteriae	severe
Sh. flexneri	intermediate
Sh. boydii	intermediate
Sh. sonnei	mild

Food Borne Infections - Vibrio

Gram negative pleomorphic bacillus (curved or straight), motile

Sources

- Primarily marine or esturine environments

Factors Affecting Growth

- Temperature range 10 - 43oC, optimum 37oC
- Typically has an NaCl requirement of 1 - 3 %
- Optimum pH range slightly alkaline (pH 7.5 - 8.5), will grow up to pH 11

Pathogenesis

- A non-invasive infection
- Colonizes mucosa and secretes a toxin
- Incubation time is dose dependent, with onset from 2 hours to4 days
- Infectious dose approximately 1000 cells

Food Borne Infections - Vibrio

- V. chlorea

- V. mimicus
- V. parahemolyticus
- V. vulnificus

Food Borne Infections - Vibrio

- Illness characterized by diarrhea, varies from mild to severe
- Bacterium colonizes mucosa of small intestine
- Secretes toxin which reverses the flow of electrolytes and water from absorbtion to secretion
- Volume can exceed 20 liters/day
- Therapy includes fluid and electrolyte replacement, tends to be self limiting
- Without treatment, death can occur in 8 - 24 hours
- Detection methods typically involve alkaline enrichment
- Media also contain NaCl
- Pre-enrichment followed by plating on thiosulfate-citrate-bilesalts-sucrose (TCBS) agar
- Biochemical and serological confirmation

Food Borne Infections - Yersinia

Gram negative pleomorphic bacillus (curved or straight), motile at <30°C, non-motile >30°C facultatively anaerobic

Sources

- Diverse, soil, water, animals (especially swine), fruits, vegetables

Factors Affecting Growth

- Temperature range -1 - 40°C, optimum 29°C will grow well at 3°C
- Optimum pH 7-8

Pathogenesis

- Not all are pathogenic;
- Virulence is plasmid mediated
- Incubation time is dose dependent, with onset from 1 -11 days duration 5 - 14 days
- Infectious dose low (few hundred cells)

Food Borne Infections - Yersinia

- Y. enterocolitica

- Y. pestis
- Y. pseudotuberculosis

Food Borne Infections - Yersinia

- Illness characterized by fever and diarrhea, can cause intense abdominal pain which mimics appendicitis
- Bacterium adheres to mucosa of small intestine by Peyer's patches
- Enters lamina propria
- Produces a local inflammatory response which disrupts fluid absorption
- Tends to be self limiting

Food Borne Infections - Yersinia

- Detection methods involve cold (4oC) enrichment
- Pre-enrichment followed by plating on cefsulodin-irgasan- novobiocin (CIN) agar
- Biochemical confirmation; congo red assay for plasmid

6

Food Fermentations: Role of Microorganisms in Food

BASIC PRINCIPLES OF FERMENTATION

THE DIVERSITY OF FERMENTED FOODS

Numerous fermented foods are consumed around the world. Each nation has its own types of fermented food, representing the staple diet and the raw ingredients available in that particular place. Although the products are well know to the individual, they may not be associated with fermentation. Indeed, it is likely that the methods of producing many of the worlds fermented foods are unknown and came about by chance. Some of the more obvious fermented fruit and vegetable products are the alcoholic beverages - beers and wines. However, several more fermented fruit and vegetable products arise from lactic acid fermentation and are extremely important in meeting the nutritional requirements of a large proportion of the worlds population. Table contains examples of fermented fruit and vegetable products from around the world.

Organisms Responsible for Food Fermentations

The most common groups of micro-organisms involved in food fermentations *are*:

- Bacteria
- Yeasts
- Moulds

Bacteria

Several bacterial families are present in foods, the majority of which are concerned with food spoilage. As a result, the important role of bacteria in the fermentation of foods is often overlooked. The most important bacteria in desirable food fermentations are the *lactobacillaceae* which have the ability to produce lactic acid from carbohydrates. Other important bacteria, especially in the fermentation of fruits and vegetables, are the acetic acid producing *acetobacter* species.

Yeasts

Yeasts and yeast-like fungi are widely distributed in nature. They are present in orchards and vineyards, in the air, the soil and in the intestinal tract of animals. Like bacteria and moulds, yeasts can have beneficial and non-beneficial effects in foods. The most beneficial yeasts in terms of desirable food fermentation are from the *Saccharomyces* family, especially *S. cerevisiae.* Yeasts are unicellular organisms that reproduce asexually by budding. In general, yeasts are larger than most bacteria. Yeasts play an important role in the food industry as they produce enzymes that favour desirable chemical reactions such as the leavening of bread and the production of alcohol and invert sugar.

Table. Fermented Foods from Around the World

Name and Region	Type of product
Indian Sub-continent	
Acar, Achar, Tandal achar, Garam nimboo achar	Pickled fruit and vegetables
Gundruk Lemon pickle , Lime pickle, Mango pickle	Fermented dried vegetable
South East Asia	
Asinan, Burong mangga, Dalok, Jeruk, Kiam-chai, Kiam-cheyi, Kongchai, Naw-mai-dong, Pak-siamdong, Sambal Paw-tsay, Phak-dong, Phonlami-dong, Sajur asin, tempo-jak, Santol, Si-sek-chai, Sunki,	Pickled fruit and vegetables
Tang-chai, Tempoyak, Vanilla, Bai-ming, Leppet-so,	Fermented tea leaves
Miang Nata de coco, Nata de pina	Fermented fruit juice
East Asia	
Bossam-kimchi, Chonggak-kimchi, Dan moogi, Dongchimi, kigactuki, Kakduggi, Kimchi, Mootsanji, Muchung-kimchi, Oigee, Oiji, Oiso baegi, Tongbaechu- kimchi, Tongkimchi, Totkal kimchi,	Fermented in brine Kachdoo
Cha-ts'ai, Hiroshimana, Jangagee, Nara senkei, Narazuke, Nozawana, Nukamiso-zuke, Omizuke, Powtsai, Red in snow, Seokbakji, Shiozuke, Szechwan cabbage, Tai-tan tsoi, Takana, Takuan, Tsa Tzai, Tsu, Umeboshi, Wasabizuke,	Pickled fruit and vegetables

Yen tsai Hot pepper sauce	
Africa	
Fruit vinegar Hot pepper sauce	Vinegar
Lamoun makbouss, Mauoloh, Msir, Mslalla, Olive	Pickled fruit and vegetables
Oilseeds, Ogili, Ogiri, Hibiscus seed seeds	Fermented fruit and vegetable
Wines	Fermented fruits
Americas	
Cucumber pickles, Dill pickles, Olives, Sauerkraut,	Pickled fruit and vegetables
Lupin seed, Oilseeds,	Pickled oilseed
Vanilla, Wines	Fermented fruit and vegetable
Middle East	
Kushuk	Fermented fruit and vegetables
Lamoun makbouss, Mekhalel, Olives, Torshi, Tursu	Pickled fruit and vegetables Wines
Fermented fruits	
Europe and World	
Mushrooms, Yeast	Moulds
Olives, Sauerkohl, Sauerruben	Pickled fruit and vegetables
Grape vinegar, Wine vinegar	Vinegar
Wines, Citron	Fermented fruits

Moulds

Moulds are also important organisms in the food industry, both as spoilers and preservers of foods. Certain moulds produce undesirable toxins and contribute to the spoilage of foods. The *Aspergillus* species are often responsible for undesirable changes in foods. These moulds are frequently found in foods and can tolerate high concentrations of salt and sugar.

However, others impart characteristic flavours to foods and others produce enzymes, such as amylase for bread making. Moulds from the genus *Penicillium* are associated with the ripening and flavour of cheeses. Moulds are aerobic and therefore require oxygen for growth. They also have the greatest array of enzymes, and can colonise and grow on most types of food. Moulds do not play a significant role in the desirable fermentation of fruit and vegetable products.

When micro-organisms metabolise and grow they release by-products. In food fermentations the by-products play a beneficial role in preserving and changing the texture and flavour of the food substrate. For example, acetic acid is the by-product of the fermentations of some fruits. This acid not only affects the flavour of the final product, but more importantly has a preservative effect on the food. For food fermentations, micro-organisms are often classified according to these by-products. The fermentation of milk to yoghurt involves a specific group of bacteria called the lactic acid bacteria (*Lactobacillus* species). This is a general name attributed to those bacteria which produce lactic acid as they grow. Acidic foods are less susceptible to spoilage than neutral or alkaline foods and hence the acid helps to preserve the product. Fermentations also result in a change in texture. In the case of milk, the acid causes the precipitation of milk protein to a solid curd.

Enzymes

The changes that occur during fermentation of foods are the result of enzymic activity. Enzymes are complex proteins produced by living cells to carry out specific biochemical reactions. They are known as catalysts since their role is to initiate and control reactions, rather than being used in a reaction. Because they are proteinaceous in nature, they are sensitive to fluctuations in temperature, pH, moisture content, ionic strength and concentrations of substrate and inhibitors. Each enzyme has requirements at which it will operate most efficiently. Extremes of temperature and pH will denature the protein and destroy enzyme activity. Because they are so sensitive, enzymic reactions can easily be controlled by slight adjustments to temperature, pH or other reaction conditions. In the food industry, enzymes have several roles - the liquefaction and saccharification of starch, the conversion of sugars and the modification of proteins. Microbial enzymes play a role in the fermentation of fruits and vegetables.

Nearly all food fermentations are the result of more than one micro-organism, either working together or in a sequence. For example, vinegar production is a joint effort between yeast and acetic acid forming bacteria. The yeast convert sugars to alcohol, which is the substrate required by the *acetobacter* to produce acetic acid. Bacteria from different species and the various micro-organisms - yeast and moulds -all have their own preferences for growing conditions, which are set within narrow limits. There are very few pure culture fermentations. An organism that initiates fermentation will grow there until it's by-products inhibit further growth and activity. During this initial growth period, other organisms develop which are ready to take over when the conditions become intolerable for the former ones.

In general, growth will be initiated by bacteria, followed by yeasts and then moulds. There are definite reasons for this type of sequence. The smaller micro-organisms are the ones that multiply and take up nutrients from the surrounding area most rapidly. Bacteria are the smallest of micro-organisms, followed by yeasts and moulds. The smaller bacteria, such as *Leuconostoc* and *Streptococcus* grow and ferment more rapidly than their close relations and are therefore often the first species to colonise a substrate

Desirable Fermentation

It is essential with any fermentation to ensure that only the desired bacteria, yeasts or moulds start to multiply and grow on the substrate. This has the effect of suppressing other micro-organisms which may be either pathogenic and cause food poisoning or will generally spoil the fermentation process, resulting in an end-product which is neither expected or desired. An everyday example used to illustrate this point is the differences in spoilage between pasteurised and unpasteurised milk.

Unpasteurised milk will spoil naturally to produce a sour tasting product which can be used in baking to improve the texture of certain breads. Pasteurised milk, however, spoils (non-desirable fermentation) to produce an unpleasant product which has to be disposed of. The reason for this difference is that pasteurisation (despite being a very important process to destroy pathogenic micro-organisms) changes the micro-organism environment and if pasteurised milk is kept unrefrigerated for some time, undesirable micro- organisms start to grow and multiply before the desirable ones.

In the case of unpasteurised milk, the non-pathogenic lactic acid bacteria start to grow and multiply at a greater rate that any pathogenic bacteria. Not only do the larger numbers of lactic acid bacteria compete more successfully for the available nutrients, but as they grow they produce lactic acid which increases the acidity of the substrate and further suppresses the bacteria which cannot tolerate an acid environment.

Most food spoilage organisms cannot survive in either alcoholic or acidic environments. Therefore, the production of both these end products can prevent a food from spoilage and extend the shelf life. Alcoholic and acidic fermentations generally offer cost effective methods of preserving food for people in developing countries, where more sophisticated means of preservation are unaffordable and therefore not an option.

The principles of microbial action are identical both in the use of micro-organisms in food preservation, such as through desirable fermentations, and also as agents of destruction via food spoilage. The type of organisms present and the environmental conditions will determine the nature of the reaction and the ultimate products. By manipulating the external reaction conditions, microbial reactions can be controlled to produce desirable results. There are several means of altering the reaction environment to encourage the growth of desirable organisms. These are discussed below.

Manipulation of Microbial Growth and Activity

There are six major factors that influence the growth and activity of micro- organisms in foods. These are moisture, oxygen concentration, temperature, nutrients, pH and inhibitors (Mountney and Gould, 1988). The food supply available to the micro-organisms depends on the composition of the food on which they grow. All micro-organisms differ in their ability to metabolise proteins, carbohydrates and fats. Obviously, by manipulating any of these six factors, the activity of micro-organisms within foods can be controlled.

Moisture

Water is essential for the growth and metabolism of all cells. If it is reduced or removed, cellular activity is decreased. For example, the removal of water from cells by drying or the change in state of water (from liquid to solid) affected by freezing, reduces the availability of water to cells (including microbial cells) for metabolic activity. The form in which water exists within the food is important as far as microbial activity is concerned. There are two types of water - free and bound.

Bound water is present within the tissue and is vital to all the physiological processes within the cell. Free water exists in and around the tissues and can be removed from cells without seriously interfering with the vital processes. Free water is essential for the survival and activity of micro-organisms. Therefore, by removing free water, the level of microbial activity can be controlled.

The amount of water available for micro-organisms is referred to as the water activity (aw). Pure water has a water activity of 1.0. Bacteria require more water than yeasts, which require more water than moulds to carry out their metabolic activities. Almost all microbial activity is inhibited below aw of 0.6. Most fungi are inhibited below aw of 0.7, most yeasts are inhibited below aw of 0.8 and most bacteria below aw 0.9.

Naturally, there are exceptions to these guidelines and several species of micro-organism can exist outside the stated range. The water activity of foods can be changed by altering the amount of free water available. There are several ways to achieve this – drying to remove water; freezing to change the state of water from liquid to solid; increasing or decreasing the concentration of solutes by adding salt or sugar or other hydrophylic compounds (salt and sugar are the two common additives used for food preservation).

Addition of salt or sugar to a food will bind free water and so decrease the aw. Alternatively, decreasing the concentration will increase the amount of free water and in turn the aw. Manipulation of the aw in this manner can be used to encourage the growth of favourable micro-organisms and discourage the growth of spoilage ones.

Table. Water Activity for Microbial Reactions

A_w	Phenomenon	Examples
0.95	1.00Pseudomonas, Bacillus, Clostridium perfringens andsome yeasts inhibited	Highly perishable foodsFoods with 40% sucrose or 7% salt
0.90	Lower limit for bacterial Salmonella, Vibrio parahaemolyticus, Clostri- dium botulinum, Lactoba- cillus and some yeasts and fungi inhibited	Foods with 55% sucrose, 12% growth. salt.Intermediate-moisture foods (aw = 0.90-0.55)
0.85	Many yeasts inhibited	Foods with 65% sucrose, 15% salt
0.80	Lower limit for most enzyme activity and growth of most fungi. Staphylococcus aureus inhibited	Fruit syrups
0.75	Lower limit for halophilic bacteria Fruit jams	
0.70	Lower limit for growth of most xerophilic fungi	
0.65	Maximum velocity of Maillard reactions	
0.60	Lower limt for growth of or xerophilic yeasts and fungi	Dried fruits (15-20% water) osmophilic
0.55	Deoxyribose nucleic acid (DNA) becomes disordered (lower limit for life to continue)	
0.50	Dried foods (aw=0-0.55)	
0.40	Maximum oxidation velocity	
0.25	Maximum heat resistance of bacterial spores	

Oxidation-Reduction potential

Oxygen is essential to carry out metabolic activities that support all forms of life. Free atmospheric oxygen is utilised by some groups of micro-organisms, while others are able to metabolise the oxygen which is bound to other compounds such as carbohydrates. This bound oxygen is in a reduced form.

Micro-organisms can be broadly classified into two groups - aerobic and anaerobic. Aerobes grow in the presence of atmospheric oxygen while anaerobes grow in the absence of atmospheric oxygen. In the middle of these two extremes are the facultative anaerobes which can adapt to the prevailing conditions and grow in either the absence or presence of atmospheric oxygen. Microaerophilic organisms grow in the presence of reduced amounts of atmospheric oxygen. Thus, controlling the availability of free oxygen is one means of controlling microbial activity within a food. In aerobic fermentations, the amount of oxygen present is one of the limiting factors. It determines the type and amount of biological product obtained, the amount of substrate consumed and the energy released from the reaction.

Moulds do not grow well in anaerobic conditions, therefore they are not important in terms of food spoilage or beneficial fermentation, in conditions of low oxygen availability.

Temperature

Temperature affects the growth and activity of all living cells. At high temperatures, organisms are destroyed, while at low temperatures, their rate of activity is decreased or suspended. Micro-organisms can be classified into three distinct categories according to their temperature preference.

Table. Classification of Bacteria According to Temperature Requirements

	Temperature Required for Growth°C			
Type of bacteria	Minimum	optimum	maximum	General sources of bacteria
Psychrophilic	0 to 5	15 to 20	30	Water and frozen foods
Mesophilic	10 to 25	30 to 40	35 to 50	Pathogenic and non-pathogenic bacteria
Thermophilic	25 to 45	50 to 55	70 to 90	Spore forming bacteria from soil and water

Nutritional Requirements

The majority of organisms are dependent on nutrients for both energy and growth. Organisms vary in their specificity towards different substrates and usually only colonise foods which contain the substrates they require. Sources of energy vary from simple sugars to complex carbohydrates and proteins. The energy requirements of micro-organisms are very high. Limiting the amount of substrate available can check their growth.

Hydrogen ion Concentration (pH)

The pH of a substrate is a measure of the hydrogen ion concentration. A food with a pH of 4.6 or less is termed a high acid or acid food and will not permit the growth of bacterial spores. Foods with a pH above 4.6. are termed low acid and will not inhibit the growth of bacterial spores. By acidifying foods and achieving a final pH of less than 4.6, most foods are resistant to bacterial spoilage. The optimum pH for most micro-organisms is near the neutral point (pH 7.0). Certain bacteria are acid tolerant and will survive at reduced pH levels. Notable acid-tolerant bacteria include *the Lactobacillus* and *Streptococcus* species, which play a role in the fermentation of dairy and vegetable products. Moulds and yeasts are usually acid tolerant and are therefore associated with spoilage of acidic foods.

Micro-organisms vary in their optimal pH requirements for growth. Most bacteria favour conditions with a near neutral pH (7). Yeasts can grow in a pH range of 4 to 4.5 and moulds can grow from pH 2 to 8.5, but favour an acid pH. The varied pH requirements of different groups of micro-organisms is used to good effect in fermented foods where successions of micro-organisms take over from each other as the pH of the environment changes. For instance, some groups of micro-organisms ferment sugars so that the pH becomes too low for the survival of those microbes. The acidophilic micro-organisms then take over and continue the reaction. The affinity for different pH can also be used to good effect to occlude spoilage organisms.

Inhibitors

Many chemical compounds can inhibit the growth and activity of micro- organisms. They do so by preventing metabolism, denaturation of the protein or by causing physical damage to the cell. The production of substrates as part of the metabolic reaction also acts to inhibit microbial action.

Controlled Fermentation

Controlled fermentations are used to produce a range of fermented foods, including sauerkraut, pickles, olives, vinegar, dairy and other products. Controlled fermentation is a form of food preservation since it generally results in a reduction of acidity of the food, thus preventing the growth of spoilage micro-organisms. The two most common acids produced are lactic acid and acetic acid, although small amounts of other acids such as propionic, fumaric and malic acid are also formed during fermentation.

It is highly probable that the first controlled food fermentations came into existence through trial and error and a need to preserve foods for consumption later in the season. It is possible that the initial attempts at preservation involved the addition of salt or seawater. During the removal of the salt prior to consumption, the foods would pass through stages favourable to acid fermentation. Although the process worked, it is likely that the causative agents were unknown.

Today, there are numerous examples of controlled fermentation for the preservation and processing of foods. However, only a few of these have been studied in any detail - these include sauerkraut, pickles, kimchi, beer, wine and vinegar production. Although the general principles and processes for many of the fermented fruit and vegetable products are the same -relying mainly on lactic acid and acetic acid forming bacteria, yeasts and moulds, the reactions have not been quantified for each product. The reactions are usually very complex and involve a series of micro-organisms, either working together or in succession to achieve the final product.

Indigenous Fermented-Food Technologies for Small-Scale Industries

Introduction

Hunger and poverty go hand in this world where vast millions must support their families on less than US$1 per day. Protein-calorie malnutrition is common in the developing world, and it leads not only to stunting of physical growth but to retardation of brain growth and mental development. Malnutrition also leads to lower resistance to infectious disease, higher death rates, and lower productivity in the work force. Vitamin A deficiency leads to xerophthalmia and tragic blindness in at least a million children every year.

What can be done to alleviate the nutritional problems that face the developing world? The Green Revolution has resulted in a vast increase in world-wide productivity of rice and wheat and has enabled mankind to more or less continue to feed its burgeoning population until now. The Green Revolution, however, has not relieved the hunger and malnutrition of millions in the developing world. The basic problem remains essentially economic: Food is generally available if people have the money to buy it, and farmers the world over will produce more food if they can sell it for a profit. At present we have no way of improving the economic status of the millions of malnourished people, unless the world should unexpectedly decide to use the 350 billion dollars spent each year on armaments (100 billion of which is spent by the Third World countries) on improving the economic and nutritional status of humankind.

The indigenous fermented foods constitute a group of foods that are produced in homes, villages, and small cottage industries at prices within the means of a majority of the consumers in the developing world. Examination of these foods may, therefore, provide clues as to how food production and preservation can be expanded and thereby contribute to improved nutrition in the developing world in the future.

Some indigenous food fermentations such as soy sauce (shoyu), Japanese miso, Indonesian tempe, Indonesian tape ke ten, Japanese sake, Indian idli and dosai, and fish and shrimp sauces and pastes have been intensively studied to determine optimum conditions for fermentation, the essential micro- organisms, the biochemical, nutritive, and flavour/texture changes that occur during the fermentation, and possible toxicological problems that can arise.

The huge international enzyme industry today can be traced directly to the indigenous Chinese and Japanese soy-sauce and Japanese miso and sake fermentations. The monosodium glutamate (MSG) industry and the relatively recent but significant nucleotide flavour-enhancing industry also are outgrowths of soy-sauce fermentation. Soy sauce, miso, and sake have become giant, heavily commercialized industries. Fish and shrimp sauces and pastes are produced industrially in Burma, South Vietnam, Thailand, and the Philippines and on a smaller scale in other countries of SouthEast Asia. Tempe is produced commercially in Indonesia and Malaysia. It also is produced by at least 53 small factories in the United States, and it is destined to become an important meat substitute at least among the American vegetarian community, which numbers an estimated 15,000,000 at present.

Classification of Indigenous Food Fermentations

Indigenous food fermentations have been classified as follows:

- Fermentations involving proteolysis of vegetable proteins by microbial enzymes in the presence of salt and/or acid with production of amino acid and peptide mixtures with a meat-like flavour (examples are soy sauce, Indonesian kocap, miso, and Indonesian tauco);
- Fermentations involving enzymic hydrolysis of fish and shrimp or other marine animals in the presence of relatively high salt concentrations to produce meatflavoured sauces and pastes;
- Fermentations producing a meat-like texture in a cereal-grain- legume substrate by means of fungal mycelium that knits the particles together (examples are Indonesian and Malaysian tempe kedela, Indonesian oncom (ontjom) kacano. oncom tahoo. and tempe bonakrek);
- Fermentations involving the koji principle, in which microorganisms with desired enzymes are grown on a cereal-grain or legume substrate to produce koji, a crude enzyme concentrate that can be used to hydrolyse particular components in the desired fermentation (this category includes soy sauce, miso, kecap, tauco, tempe, oncom, tape, and rice wines such as sake);
- Fermentation in which organic acids are major products (this category includes Korean kimchi, sauerkraut, fermented milks and cheeses, African ogi and uji, idli, dosai, tape, and also tempe, in which acidification occurs during the initial soaking of the soybeans),
- Fermentations in which ethanol is a major product (this category includes rice wines, palm toddies, sugar cane wines, beers, and, very important nutritionally, tape ketan and tape ketela).

There is considerable overlap among the various categories of fermentation, but the classification does facilitate description and analysis of the potential usefulness of the various fermentations.

Examination of the various classifications indicates that food fermentations could be used to establish the following small-scale food-processing operations:

- Production of meat-like flavours from vegetable proteins (soy sauce and miso),
- Production of meat-like flavours from fish and shrimp and other marine animals (fish and shrimp sauces and pastes),
- Production of meat-like textures in cereal-legume substrates (tempe and oncom),
- Production of highly starchy foods with improved protein content (tape ketan and tape ketela),
- Production of foods with improved vitamin content (tape, tempe, and oncom),
- Production of leavened bread-type foods without the use of wheat or rye (idli and dosai),
- Production of quick-cooking food (tempe and oncom).

Production of Meat-like Flavours from Vegetable Proteins

Foods are generally sought after and consumed on the basis of their desirable aromas, textures, and flavours. People like foods that taste good. Meats have high prestige and are expensive, but excellent, sources of proteins and other nutrients. There is not at present enough meat in the world to supply the world's nutritional needs for protein at prices the consumer can afford to pay. Even in Australia, which is noted for its high consumption of meat, Marmite and Vegemite, acid-hydrolysed yeast products with a meatlike flavour, are widely consumed as a spread for bread and as an ingredient in soups.

In the United States, where meat dishes are the mainstay of the diet, nearly every home now has its bottle of soy sauce used to add meat flavour to various dishes. In Africa, where meats are a minor part of the diet, bouillon and Maggi cubes are consumed by the billions. There is a considerable need for meat flavours and the amino acids and peptides in meatflavoured sauces and pastes for modifying and formulating foods in the diets of people in the developing world. A small-scale industry can be based on the production of meat-like flavours.

Soy Sauce (Shoyu) and Miso Fermentations

The Chinese thousands of years ago showed the world how to produce meat-like flavours from vegetable protein, which was one of the great discoveries in food science and which may become of even greater importance in the future. The soy sauce and miso fermentations were originally household fermentations, and they continue to be so in parts of the Orient today even though the products are also manufactured industrially in very targe quantities (21, 3337).

Miso has been made in the home and on the farm since A.D. 1600 in Japan (20). Indigenous miso manufacture is a very interesting process. Soybeans are soaked, cooked, mashed, and then formed into balls. The balls are tied together with rice straw and suspended above a stove or heater for about 30 days, during which time the balls become overgrown with moulds naturally present in the environment. The balls are then brushed, mixed with water and salt, and packed into crocks where fermentation continues for a year or longer. The resulting meat-flavoured paste is a primitive miso. Free liquid is used as a soy sauce.

At present, the Japanese use a koji for the manufacture of both soy sauce and miso. For soy sauce, the koji is prepared by overgrowing soaked, cooked, cooled soybeans coated with ground roasted wheat with moulds belonging to the Aspergillus oryzee species. Production of the koji is essentially a solid-state fermentation. The soy-sauce koji contains proteases, amylases, and lipases that hydrolyse their respective substrates in the subsequent submerged fermentation in 18 per cent weight per volume salt brine. During the submerged fermentation, Pediococcus cerevisiae, Lactobacillus delbruekii, and salt-tolerant Saccharomyces rouxii develop.

For miso, the koji is prepared from rice, barley, or soybeans. The fermenting organism again is Aspergillus oryzee. Following overgrowth of the mould, the koji is mixed with hydrated, cooked soybeans and salt. Miso is essentially in a solid state during its entire fermentation. Hesseltine and Shibasaki (9) demonstrated that the only micro-organism essential for the miso fermentation (in addition to the mould Aspergillus oryzee) was Saccharomyces rouxil, which develops when the pH of the mash has fallen to below 5.0. However, miso from an earlier fermentation can be used to inoculate a new fermentation (at the time of mixing the koji with salt and soybeans).

The satt content in various misos varies from 5.5 to 13 per cent weight per volume in fresh miso (20). Sweetness depends on the proportion of cereal koji added to the soybeans (4). Colour depends in part on the total cooking time given the soybeans and the length and temperature of fermentation.

The higher the salt content, the slower the proteolytic hydrolysis and the better the miso keeps

Both soy sauce and miso are highly industrialized in Japan. Yet, by their nature, both fermentations can be conducted on a small scale at the cottage-industry level or even in the home. In the miso-dame process, the fermentation depends on moulds naturally present in the environment and on the straw used to tie the soybean-mash balls. In some places in the Orient soy-sauce manufacturers today rely on the moulds present in wheat flour to produce a soy-sauce koji.

More recently, soy sauce has been manufactured by hydrolysing soybeans with concentrated hydrochloric acids and neutralizing the product with sodium hydroxide. While this does yield a meat-like flavour, the alcohols, organic acids, and esters produced by the fermentation process are lacking and the acid-hydrolysed sauce does not have the aroma and flavour of the genuine fermented sauce. Also, tryptophan is destroyed during acid hydrolysis, resulting in a loss of nutritive value.

The soy-sauce-miso process is used to produce meat-like flavours from soybeans, wheat, rice, and barley substrates. It could probably be adapted to other substrates, such as coconut and yeast, which yield meat-like flavours by acid hydrolysis. These meat-flavoured sauces and pastes could improve the nutrition and diets in many developing countries. The demand would support small-scale factories in most countries.

Fish and Shrimp Sauces and Pastes

While soy sauce was developed in northern Asia, the South East Asians made an equally great discovery concerning how to convert small, surplus fish and shrimp into meat-flavoured sauces and pastes. In fact, the soy-sauce and fish sauce fermentations both depend on proteolytic enzymes to hydrolyse the proteins in the substrate to the constituent amino acids and peptides. In soy sauce, fungal enzymes are used, while in fish sauce, enzymes in the fish tissues, particularly the gut tissues, are involved.

Both fermentations are carried out in concentrated salt brine (18 per cent for soy sauce; 23 per cent or higher for fish sauce). In their highest qualities, soy sauce and fish sauce are quite comparable in flavour: both are similar to beef broths. The processes are amenable to the use of surplus marine animals of many types (non-commercial) and could serve as a base for small-scale factories.

Over the past few years a number of accelerated processes for manufacturing fish sauces and pastes have been developed. Some of these involve the addition of vegetable proteases such as bromeli or papain to the fermenting fish .Ismail developed a "rapid" process for fish sauce in which he produced a koji by growing selected strains of Aspergillus oryzee on soybeans and then mixing the koji with an equal weight of fish, thus effectively producing a fish-soy sauce.

Many new products are technologically possible by combining known fermentations with new substrates or new microorganisms. Much of the world has not as yet adopted fishshrimp sauces or pastes to its diet, although they present an excellent opportunity to utilize surplus marine animals as substrates. The processes offer an opportunity to entrepreneurs who wish to start new businesses wherever surplus marine animals are available.

Meat Substitutes (Analogues)

Large Western food companies have invested millions of dollars in developing processes by which soybean protein is extracted and concentrated to purities above 90 per cent and then by chemical modification and extrusion through platinum dies spun into protein strands which can be formed into pieces with meat-like texture. With added fats and meat flavors, the products are called meat analogues (a sophisticated name for imitation meats) and there is no doubt that these vegetable protein products will be an important part of "meat" consumption in the Western world in the future. Several products are already on the market .

Similarly, large meat packers have developed processes in which soybeans are flaked, tempered, formulated, and extruded so that the products are subjected to high pressure and temperature for a short time and emerge from the extruder as chewy, protein-rich, meat-like nuggets that supply the flavour, texture, and nutritive value of meats in a number of Western dishes. In England, Rank, Hovis, and McDougall developed a process wherein mould mycelium is grown on low-cost carbohydrate, recovered by filtration, and formulated with added fat, flavour, and other components to produce meat analogues in which the mould mycelium provides the fibrous texture .This process entails modern sophisticated food science and technology. It is interesting that the Western world, which consumes so much meat, has developed these advanced processes for manufacturing vegetable-protein meat substitutes.

Indonesian Tempe Kedele

The Indonesians centuries ago, without modern chemistry and microbiology, developed a fermentation process for making meat analogues from soybeans in which the mould mycelium provides the meat-like texture.

The product, called tempe kedele, is manufactured by small-scale cottage industries, which could be established in any developing country where soybeans are available.

Tempe kedele is a white, mould-covered cake produced by fungal fermentation of dehulled, hydrated (soaked), and partially cooked soybean cotyledons. The mould grows throughout the bean mass, knitting it into a compact cake that can be sliced thin and deep-fat fried or cut into chunks and used as a protein-rich meat substitute in soups. The essential moulds belong to the genus Rhizoous; Rhizoous oligosporus is the species identified as most characteristic and best adapted for production of tempe .

The essential steps in the production of tempe are the following:

- Cleaning the soybeans,
- Hydration and acid fermentation,
- Dehulling dry or following hydration,
- Partial cooking,
- Draining, cooling, surface drying,
- Placing the soybean cotyledons in suitable fermentation containers,
- Inoculating with tempe mould (before or after placing in the fermentation container),
- Incubating until the cotyledons are completely covered with mould mycelium,
- Harvesting and selling,
- Cooking for consumption, either by frying in deep fat or by use as an ingredient in soups in place of meat.

Traditional Tempe Fermentation

The soybeans are washed and soaked in water overnight, during which time they undergo bacterial acid fermentation reducing the pH to 5.0 or lower. An alternate process is to place the soybeans in water, bring it to a boil, and then allow the beans to soak overnight. The general purpose of the boiling is to facilitate removal of the hulls. The hulls are loosened by rubbing the soaked beans between the hands or stamping them with the feet and are then floated away with water. The cotyledons are then given a short boil, cooled, surface- dried by winnowing, and inoculated with tempe mould either from a previous batch of sound tempe or from mould grown and dried on leaves. Traditionally, the inoculated cotyledons are then wrapped in small packets using wilted banana or other large leaves and are incubated in a warm place for two or three days, during which time they are competely overgrown by the mould mycelium. The tempe is then ready for cooking.

Industrial Production of Tempe

Twenty years ago most tempe was prepared for sale using the traditional process described above. Then research in the United States resulted in some improvements in medium-scale processing of tempe. Steinkraus et al. described a pilot-plant process in which the soybeans were dehulled dry by passing them through a properly adjusted burr mill. Preceding the burr mill, the beans were given a short heat treatment at 104°C (220°F) to shrivel the cotyledons.

The hulls were then removed from the cotyledons by passing them through an aspirator or over an Oliver gravity separator. Alternatively, the beans were soaked and dehulled wet by passing them through an abrasive vegetable peeler. Acidification of the beans, considered to be an essential step, particularly in large-scale processing, where invasion by foodspoilage organisms could ruin large batches, was accomplished by adding 1 per cent weight per volume of lactic acid to the soak and cook water. The partially cooked beans were then drained, cooled, and inoculated with powdered, pure- culture tempe mould grown on sterilized soybeans and freeze-dried.

The inoculum was mixed with the drained, cooled cotyledons in a Hobart mixer. The inoculated beans were then spread on dryer trays (35 x 81 x 1.3 cm), covered with a layer of wax paper, and incubated at 37°C (98.6°F) and 90 per cent relative humidity. By this procedure, fermentation was complete in less than 24 hours. The tempe was cut into 2.5 cm squares, and the dryer trays were placed in a circulatinghot-air dryer at 104°C (220 F), dehydrated to less than 10 per cent moisture, and packaged in polyethylene bags for distribution. Within a few years, the commercial tempe industry in Indonesia had adopted wooden trays with dimensions similar to those used above. They lined the trays with plastic sheeting perforated to allow access of air to the mould.

Martinelli and Hesseltine (16) developed a new method of incubating the tempe in plastic bags with perforations at 0.25 to 1.3-cm intervals to allow access to oxygen. In this method the soybean cotyledons are inoculated with the mould and placed in the plastic bags or in plastic tubes similar to sausage casings. They can be incubated immediately or stored in a refrigerator until fermentation is desired. Then the mould overgrows the soybeans in a day or less. The plastic-bag process has been widely adopted in Indonesia and is also being used commercially in new tempe factories in the United States.

According to Shurtleff and Aoyagi (22), there are 53 tempe factories in the United States, the largest of which produces 7,000 pounds of tempe per day; the largest operation in Indonesia produces 1,760 pounds of tempe per day (20). Thus, tempe production is still a relatively small commercial operation. It offers many opportunities for new factories in countries where protein-rich meat substitutes at relatively low cost are needed by the consumers.

During tempe fermentation, not only is texture introduced into the soybean cotyledons: the proteins are partially hydrolysed; the lipids are hydrolysed to their constituent free fatty acids; stachyose, a tetrasaccharide indigestible by human beings, is reduced; riboflavin is nearly doubled; and niacin increases seven times.

Also, vitamin B12, usually lacking in vegetarian foods, is synthesized by a bacterium that grows along with the essential mould (15). The bacterium has been identified as Klebsiella pneumoniae (a non-pathogenic strain) (3). Thus, tempe with its high protein content (about 40 per cent on a dry basis) can supply the consumer not only with the essential protein requirements but also with the requirements for vitamin B12

The tempe fermentation has been applied to a wide variety of bean types (5). A new type of tempe in which wheat and soybeans are combined was developed by Wang and Hesseltine (10, 31). The tempe process is a resource of considerable potential industrial value as it can be used to introduce meat-like textures into cereal-legume substrates. It also decreases the total cooking time required for soybeans from about five or six hours to about 40 minutes boiling (30 minutes precooking and 10 minutes following fermentation). It is one of the world's first quick-cooking foods-a quality highly prized in modern food science.

Tempe has been rapidly adopted by American vegetarians and is becoming increasingly available in health-food stores and even large supermarkets in the eastern and western United States. Tempeburgers are available on the West Coast and will likely become a staple in the diet. It will probably become an accepted food in developing countries at present unfamiliar with its desirable characteristics.

A Process for Raising the Protein Content of high-starch Substrates Millions of people in the world today, particularly the poor, use cassava (manioc, yuca) as a stapte in the diet. It is an excellent source of calories but is entirely too low in protein to provide the needs of the consumer. The Indonesians centuries ago developed a fermentation process whereby the protein content of cassava or any other substrate rich in starch can be increased, improving its overall nutritive value. The product is called tape ketan when rice is the substrate and tape ketela when cassava is the substrate.

The essential micro-organisms are Amylomyces rouxli and a yeast of the Endomycoosis burtonii type (2, 14). The organisms are available in the markets of most South East Asian countries as a ragi cake. Glutinous rice is steamed, cooled, and inoculated with powdered ragi. Incubated in a warm place, the rice becomes a delicious sweet/sour alcoholic food within two or three days. The micro-organisms utilize a portion of the starch, and the protein content of the tape ketan reaches approximately 16 per cent (dry basis) - about double the initial protein content of the rice. Of considerable nutritional importance, Iysine-the first limiting amino acid in rice-is selectively synthesized, increasing about 15 per cent, and thiamine, which is very low in polished rice, increases 300 per cent, reaching a level close to that found in unpolished rice.

If cassava is the substrate, the tubers are peeled and steamed and inoculated with the powdered ragi cake. The tubers become sweetlsour and alcholic. Again a portion of the starch is utilized, and the protein content on a dry basis can reach as high as 3 per cent, or even higher if the microorganisms use a higher proportion of the starch in the tuber. The product can be sun-dried and used as an ingredient in soups.

The tape fermentation, which provides a means of raising the protein content of high starch substrates and also increases the Iysine and thiamine content of starchy substrafes, is a potentially valuable industrial resource. It should be noted that tape ketan would very likely be a highly acceptable new food in many developing countries if it were produced by small factories and distributed in an attractive preserved form.

Leavened Bread-like Foods without use of Wheat or Rye

Wheat and rye breads are staples in much of the Western world, and there has been considerable interest in developing leavened breads in countries where wheat or rye flours are unavailable. Moreover, the Indians long ago developed a process whereby leavened breads, called idli when steamed and dosai when fried, can be made from rice and black-gram dahl or other legumes. The process is a home industry.

The housewife soaks rice and black-gram dahl separately during the day. In the evening she grinds them separately in a mortar and pestle; the rice is coarsely ground and the dahl finely ground for idli, and both are finely ground for dosai. She then combines them with water and a small amount of salt for seasoning to make a thick batter. This mixture is incubated overnight, during which time Leuconostoc mesenteroides grows, producing both lactic acid, which lowers the pH, and carbon dioxide, which leavens the batter.

The batter is then steamed in the form of small white cakes in the morning or fried as a pancake. Idli is a type of sourdough bread. Soybeans can be substituted for the black-gram dahl. It is likely that other cereal grains and perhaps even cassava could be substituted for the rice (17, 29). Idli-like foods are low-cost and could contribute to adequate nutrition in many developing countries if they were manufactured either in the home or in small factories.

Summary

The developing world is rich in indigenous food fermentations that can contribute significantly to world small-scale food processing and consumption over the next 20 to 50 years as population reaches six to eight billion. The world needs low-cost methods of providing nutritious protein rich meat analogues for its millions of consumers. The Indonesian tempe fermentation can serve as a model. A bacterium present in commercial tempe can be used to add vitamin B12 to other vegetarian foods.

Fuel requirements for cooking can be decreased by applying a fungal fermentation of the tempe/ontjom type to legume substrates. The world needs high-quality meat flavours derived from vegetable protein. The soy-sauce (kecap)/miso (tauco) processes and the fishlshrimp sauce and paste processes can be modified to yield a wide variety of meat-like flavours for use in formulating new foods. The protein content of high-starch substrates can be increased by applying the Indonesian tape fermentation. Leavened sourdough bread like products can be produced without the use of wheat or rye flours, using the Indian idli-dosai fermentation process. Production of such foods by small-scale food processors will contribute both to the economy of the country and to the nutritional improvement of consumers.

FERMENTED AND MICROBIAL FOOD

BACTERIAL FERMENTATIONS

What are Bacteria

Bacteria are "a large group of unicellular or multi-cellular organisms lacking chlorophyll, with a simple nucleus, multiplying rapidly by simple fission, some species developing a highly resistant resting (spore) phase; some species reproduce sexually, and some are motile. In shape they are spherical, rodlike, spiral, or filamentous. They occur in air, water, soil, rotting organic material, animals and plants. Saprophytic forms are more numerous than parasites. A few forms are autotrophic" (Walker, 1988). There are several bacterial families present in foods, the majority of which are concerned with food spoilage. The important role of bacteria in the fermentation of foods is often overlooked.

Lactic Acid Bacteria

The lactic acid bacteria are a group of Gram positive bacteria, non-respiring, non-spore forming, cocci or rods, which produce lactic acid as the major end product of the fermentation of carbohydrates. They are the most important bacteria in desirable food fermentations, being responsible for the fermentation of sour dough bread, sorghum beer, all fermented milks, cassava (to produce *gari* and *fufu*) and most "pickled" (fermented) vegetables. Historically, bacteria from the genera *Lactobacillus, Leuconostoc, Pediococcus* and *Streptococcus* are the main species involved. Several more have been identified, but play a minor role in lactic fermentations. Lactic acid bacteria were recently reviewed by Axelsson (1998).

Lactic acid bacteria carry out their reactions - the conversion of carbohydrate to lactic acid plus carbon dioxide and other organic acids - without the need for oxygen. They are described as microaerophilic as they do not utilise oxygen. Because of this, the changes that they effect do not cause drastic changes in the composition of the food. Some of the family are homofermentative, that is they only produce lactic acid, while others are heterofermentative and produce lactic acid plus other volatile compounds and small amounts of alcohol. *Lactobacillus acidophilus, L. bulgaricus, L. plantarum, L. caret, L. pentoaceticus, L brevis* and *L. thermophilus* are examples of lactic acid-producing bacteria involved in food fermentations.

All species of lactic acid bacteria have their own particular reactions and niches, but overall, *L. plantarum* – a homofermenter -produces high acidity in all vegetable fermentations and plays the major role. All lactic acid producers are non-motile gram positive rods that need complex carbohydrate substrates as a source of energy. The lactic acid they produce is effective in inhibiting the growth of other bacteria that may decompose or spoil the food.

Because the whole group are referred to as 'lactic acid bacteria' it might appear that the reactions they carry out are very simple, with the production of one substrate. This is far from the truth. The lactic acid bacteria are a diverse group of organisms with a diverse metabolic capacity. This diversity makes them very adaptable to a range of conditions and is largely responsible for their success in acid food fermentations.

Despite their complexity, the whole basis of lactic acid fermentation centres on the ability of lactic acid bacteria to produce acid, which then inhibits the growth of other non-desirable organisms. All lactic acid producers are micro- aerophilic, that is they require small amounts of oxygen to function. Species of the genera *Streptococcus* and *Leuconostoc* produce the least acid. Next are the heterofermentative species of *Lactobacillus* which produce intermediate amounts of acid, followed by the *Pediococcus* and lastly the homofermenters of the *Lactobacillus* species, which produce the most acid.

Homofermenters, convert sugars primarily to lactic acid, while heterofermenters produce about 50% lactic acid plus 25% acetic acid and ethyl alcohol and 25% carbon dioxide. These other compounds are important as they impart particular tastes and aromas to the final product. The heterofermentative lactobacilli produce mannitol and some species also produce dextran. *Leuconostoc mesenteroides* is a bacterium associated with the sauerkraut and pickle fermentations. This organism initiates the desirable lactic acid fermentation in these products. It differs from other lactic acid species in that it can tolerate fairly high concentrations of salt and sugar (up to 50% sugar). *L. mesenteroides* initiates growth in vegetables more rapidly over a range of temperatures and salt concentrations than any other lactic acid bacteria. It produces carbon dioxide and acids which rapidly lower the pH and inhibit the development of undesirable micro-organisms.

The carbon dioxide produced replaces the oxygen, making the environment anaerobic and suitable for the growth of subsequent species of lactobacillus. Removal of oxygen also helps to preserve the colour of vegetables and stabilises any ascorbic acid that is present.

Organisms from the gram positive *Propionibacteriaceae* family are responsible for the flavour and texture of some fermented foods, especially Swiss cheese, where they are responsible for the formation of 'eyes' or holes in the cheese.

These bacteria break down lactic acid into acetic and propionic acids and carbon dioxide. Several other bacteria, for *instance Leuconostoc citrovorum L. Dextranicum, Streptococcus lactis, S. Cremis, & liquefaciens* and *Brevibacterium* species are important in the fermentation of dairy products.

Lactic acid Fermentation

The lactic acid bacteria belong to two main groups – the homofermenters and the heterofermenters. The pathways of lactic acid production differ for the two. Homofermenters produce mainly lactic acid, via the glycolytic (Embden–Meyerhof) pathway).

Heterofermenters produce lactic acid plus appreciable amounts of ethanol, acetate and carbon dioxide, via the 6-phosphoglucanate/phosphoketolase pathway. The glycolytic pathway is used by all lactic acid bacteria except leuconostocs, group III lactobacilli, oenococci and weissellas.

Normal conditions required for this pathway are excess sugar and limited oxygen. Axelsson (1998) gives an in-depth account of the biochemical pathways for both homo- and hetero-fermenters.

- Homolactic fermentation: The fermentation of 1 mole of glucose yields two moles of lactic acid;

  $C_6H_{12}O_6 \rightarrow 2CH_3CHOHCOOH$

  Glucose lactic acid

- Heterolactic Fermentation: The fermentation of 1 mole of glucose yields 1 mole each of lactic acid, ethanol and carbon dioxide;

  $C_6H_{12}O_6 \rightarrow CH_3CHOHCOOH + C_2H_5OH + CO_2$

  Glucose lactic acid+ ethanol + carbon dioxide

Table. Major Lactic Acid Bacteria in Fermented Plant Products

Homofermenter	Facultative Homofermenter	Obligate Heterofermenter
Enterococcus faecium	Lactobacillus bavaricus	Lactobacillus brevis
Enterococcus faecalis	Lactobacillus casei	Lactobacillus buchneri
Lactobacillus acidophilus	Lactobacillus coryniformis	Lactobacillus cellobiosus
Lactobacillus lactis	Lactobacillus curvatus	Lactobacillus confusus
Lactobacillus delbrueckii	Lactobacillus plantarum	Lactobacillus coprophilus
Lactobacillusleichmannii	Lactobacillus sake	Lactobacillus fermentatum
Lactobacillus salivarius	Lactobacillus sanfrancisco	
Streptococcus bovis		Leuconostoc dextranicum
Streptococcus thermophilus		Leuconostoc mesenteroides
Pediococcus acidilactici		Leuconostoc paramesenteroides
Pedicoccus damnosus		Pediococcus pentocacus

Acetic Acid Bacteria

A second group of bacteria of importance in food fermentations are the acetic acid producers from the *Acetobacter* species. *Acetobacter* are important in the production of vinegar (acetic acid) from fruit juices and alcohols. The same reaction also occurs in wines, oxygen permitting, where the *acetobacter* can cause undesirable changes – the oxidation of alcohol to acetic acid. This produces a vinegary off-taste in the wine.

The most desirable action of acetic acid bacteria is in the production of vinegar. The vinegar process is essentially a two stage process, where yeasts convert sugars into alcohol, followed by *acetobacter*, which oxidise alcohol to acetic acid.

Acetic acid Fermentation

Acetobacter convert alcohol to acetic acid in the presence of excess oxygen. Oxidation of alcohol to acetic acid and water The oxidation of one mole of ethanol yields one mole each of acetic acid and water;

$$\underset{\text{Alcohol}}{C_2H_5OH} + O_2 \rightarrow \underset{\text{acetic acid}}{CH_3COOH} + \underset{\text{water}}{H_2O}$$

Bacteria of Alkaline Fermentations

A third group of bacteria are those which bring about alkaline fermentations - the *Bacillus* species. Of note are *Bacillus subtilis, B. licheniformis* and *B. pumilius. Bacillus subtilis* is the dominant species, causing the hydrolysis of protein to amino acids and peptides and releasing ammonia, which increases the alkalinity and makes the substrate unsuitable for the growth of spoilage organisms.

Alkaline fermentations are more common with protein rich foods such as soybeans and other legumes, although there are a few examples utilising plant seeds. For example water melon seeds (*Ogiri* in Nigeria) and sesame seeds (*Ogiri-saro* in Sierra Leone) and others where coconut and leaf proteins are the substrates (Indonesian *semayi* and Sudanese *kawal* respectively).

Although the range of products of alkaline fermentation does not match those brought about by acid fermentations, they are important in that they provide protein rich, low cost condiments from leaves, seeds and beans, which contribute to the diet of millions of people in Africa and Asia. Steinkraus presents a comprehensive review of the acid, alkaline and alcoholic fermentations from around the world.

Conditions Required for Bacterial Fermentations

Micro-organisms vary in their optimal pH requirements for growth. Most bacteria favour conditions with a near neutral pH (7). The varied pH requirements of different groups of micro-organisms is used to good effect in fermented foods where successions of micro-organisms take over from each other as the pH of the environment changes. Certain bacteria are acid tolerant and will survive at reduced pH levels. Notable acid-tolerant bacteria include *the Lactobacillus* and *Streptococcus* species, which play a role in the fermentation of dairy and vegetable products. Oxygen requirements vary from species to species. The lactic acid bacteria are described as microaerophilic as they carry out their reactions with very little oxygen. The acetic acid bacteria however, require oxygen to oxidise alcohol to acetic acid. In vinegar production, oxygen has to be made available for the production of acetic acid, whereas with wine it is essential to exclude oxygen to prevent oxidation of the alcohol and spoilage of the wine.

Temperature

Different bacteria can tolerate different temperatures, which provides enormous scope for a range of fermentations. While most bacteria have a temperature optimum of between 20 to 30ºC, there are some (the thermophiles) which prefer higher temperatures (50 to 55ºC) and those with colder temperature optima (15 to 20ºC). Most lactic acid bacteria work best at temperatures of 18 to 22ºC. The *Leuconostoc* species which initiate fermentation have an optimum of 18 to 22ºC. Temperatures above 22ºC, favour the *lactobacillus* species.

Salt Concentration

Lactic acid bacteria tolerate high salt concentrations. The salt tolerance gives them an advantage over other less tolerant species and allows the lactic acid fermenters to begin metabolism, which produces acid, which further inhibits the growth of non-desirable organisms. Leuconostoc is noted for its high salt tolerance and for this reason, initiates the majority of lactic acid fermentations.

Water Activity

In general, bacteria require a fairly high water activity (0.9 or higher) to survive. There are a few species which can tolerate water activities lower than this, but usually the yeasts and fungi will predominate on foods with a lower water activity.

Hydrogen ion concentration (pH)

The optimum pH for most bacteria is near the neutral point (pH 7.0). Certain bacteria are acid tolerant and will survive at reduced pH levels. Notable acid-tolerant bacteria include *the Lactobacillus* and *Streptococcus* species, which play a role in the fermentation of dairy and vegetable products.

Oxygen Availability

Some of the fermentative bacteria are anaerobes, while others require oxygen for their metabolic activities. Some, lactobacilli in particular, are microaerophilic. That is they grow in the presence of reduced amounts of atmospheric oxygen. In aerobic fermentations, the amount of oxygen present is one of the limiting factors. It determines the type and amount of biological product obtained, the amount of substrate consumed and the energy released from the reaction. Acetobacter require oxygen for the oxidation of alcohol to acetic acid.

Nutrients

All bacteria require a source of nutrients for metabolism. The fermentative bacteria require carbohydrates – either simple sugars such as glucose and fructose or complex carbohydrates such as starch or cellulose. The energy requirements of micro-organisms are very high. Limiting the amount of substrate available can check their growth.

Principles of lactic acid Fermentation

Sauerkraut is one example of an acid fermentation of vegetables. The name sauerkraut literally translates as acid cabbage. The 'sauerkraut process' can be applied to any other suitable type of vegetable product. Because of the importance of this product in the German diet, the process has received substantial research in order to commercialise and standardise production. As a result, the process and the contributing micro-organisms are known intimately. Other less well known fermented fruits and vegetables have received less research attention, therefore little is known of the exact process. It is safe to assume however that the acid fermentation of vegetables is based on this process. Lactic acid fermentations are carried out under three basic types of condition:– dry salted, brined and non-salted. Salting provides a suitable environment for lactic acid bacteria to grow which impart the acid flavour to the vegetable.

Dry salted Fermented Vegetables

With dry salting, the vegetable is treated with dry salt. The salt extracts the juice from the vegetable and creates the brine. The vegetable is prepared, washed in potable cold water and drained. For every 100 kg of vegetables 3 kg of salt is needed. The vegetables are placed in a layer of about 2.5cm depth in the fermenting container (a barrel or keg). Salt is sprinkled over the vegetables. Another layer of vegetables is added and more salt added. This is repeated until the container is three quarters full.

A cloth is placed above the vegetables and a weight added to compress the vegetables and assist the formation of a brine which takes about 24 hours. As soon as the brine is formed, fermentation starts and bubbles of carbon dioxide begin to appear. Fermentation takes between one and four weeks depending on the ambient temperature. Fermentation is complete when no more bubbles appear, after which time the pickle can be packaged in a variety of mixtures. These can be vinegar and spices or oil and spices (Lal *et al*, 1986).

The 'Sauerkraut' Process

Lactic acid bacteria are the primary group of organisms involved in sauerkraut fermentation.

They can be divided into three groups according to their types and end products:

- *Leuconostoc mesenteroides* an acid and gas producing coccus
- *Lactobacillus plantarum* and bacilli that produce acid *L. Cucumeris* and a small amount of gas
- *Lactobacillus pentoaceticus* acid and gas producing bacilli (*L. Brevis*) In addition to the desirable bacteria there are a range of undesirable micro-

Organisms present on cabbage (and other vegetable material) which can interfere with the sauerkraut process if allowed to multiply unchecked. The quality of the final product depends largely on how well the undesirable organisms are controlled during the fermentation process. Some of the typical spoilage organisms utilise the protein as an energy source, producing unpleasant odours and flavours.

The Fermentation Process

Shredded cabbage or other suitable vegetables are placed in a jar and salt is added. Mechanical pressure is applied to the cabbage to expel the juice, which contains fermentable sugars and other nutrients suitable for microbial activity. The first micro-organisms to start acting are the gas-producing cocci *(L. Mesenteroides).*

These microbes produce acids. When the acidity reaches 0.25 to 0.3% (calculated as lactic acid), these bacteria slow down and begin to die off, although their enzymes continue to function. The activity initiated by the *L. mesenteroides* is continued by the lactobacilli *(L. plantarum and L. Cucumeris)* until an acidity level of 1.5 to 2% is attained. The high salt concentration and low temperature inhibit these bacteria to some extent. Finally, *L. pentoaceticus* continues the fermentation, bringing the acidity to 2 to 2.5% thus completing the fermentation.

The end products of a normal kraut fermentation are lactic acid along with smaller amounts of acetic and propionic acids, a mixture of gases of which carbon dioxide is the principal gas, small amounts of alcohol and a mixture of aromatic esters. The acids, in combination with alcohol form esters, which contribute to the characteristic flavour of sauerkraut. The acidity helps to control the growth of spoilage and putrefactive organisms and contributes to the extended shelf life of the product. Changes in the sequence of desirable bacteria, or indeed the presence of undesirable bacteria, alter the taste and quality of the product.

Effects of Temperature on Sauerkraut Process

The optimum temperature for sauerkraut fermentation is around 21ºC. A variation of just a few degrees from this temperature alters the activity of the microbial process and affects the quality of the final product. Therefore, temperature control is one of the most important factors in the sauerkraut process. A temperature of 18º to 22º C is most desirable for initiating fermentation since this is the optimum temperature range for the growth and metabolism of *L. mesenteroides*. Temperatures above 22ºC favour the growth of *Lactobacillus* species.

Effects of Salt on the Sauerkraut Process

Salt plays an important role in initiating the sauerkraut process and affects the quality of the final product. The addition of too much salt may inhibit the desirable bacteria, although it may contribute to the firmness of the kraut. The principle function of salt is to withdraw juice from the cabbage (or other vegetable), thus making a more favourable environment for development of the desired bacteria.

Generally, salt is added to a final concentration of 2.0 to 2.5%. At this concentration, lactobacilli are slightly inhibited, but cocci are not affected. Unfortunately, this concentration of salt has a greater inhibitory effect against the desirable organisms than against those responsible for spoilage. The spoilage organisms can tolerate salt concentrations up to between 5 and 7%, therefore it is the acidic environment created by the lactobacilli that keep the spoilage bacteria at bay, rather than the addition of salt. In the manufacture of sauerkraut, dry salt is added at the rate if 1 to 1.5 kg per 50kg cabbage (2 to 3%). The use of salt brines is not recommended in sauerkraut making, but is common in vegetables that have a low water content. It is essential to use pure salt since salts with added alkali may neutralise the acid.

Use of Starter Cultures

In order to produce sauerkraut of consistent quality, starter cultures (similar to those used in the dairy industry) have been recommended. Not only do starter cultures ensure consistency between batches, they speed up the fermentation process as there is no time lag while the relevant microflora colonise the sample. Because the starter cultures used are acidic, they also inhibit the undesirable micro-organisms.

It is possible to add starters traditionally used for milk fermentation, such as *Streptococcus lactis*, without adverse effect on final quality. Because these organisms only survive for a short time (long enough to initiate the acidification process) in the kraut medium, they do not disturb the natural sequence of micro-organisms. On the other hand, if *Leuconostoc mesenteroides* is added in the early stages, it gives a good flavour to the final product, but alters the sequence of subsequent bacterial growth and results in a product that is incompletely fermented.

If gas producing rods (for example L *pentoaceticus)* are added to the sauerkraut, this disturbs the balance between acetic and lactic acids - more acetic acid and less lactic acid are produced than normal - and the fermentation never reaches completion. If lactic acid, non-gas producing *rods (L. Cucumeris)* are used as a starter, again the kraut is not completely fermented and the resulting product is bitter and more susceptible to spoilage by yeasts.

It is possible to use the juice from a previous kraut fermentation as a starter culture for subsequent fermentations. The efficacy of using old juice depends largely on the types of organisms present in the juice and its acidity. If the starter juice has an acidity of 0.3% or more, it results in a poor quality kraut. This is because the cocci which would normally initiate fermentation are suppressed by the high acidity, leaving the bacilli with sole responsibility for fermentation. If the starter juice has an acidity of 0.25% or less, the kraut produced is normal, but there do not appear to be any beneficial effects of adding this juice. Often, the use of old juice produces a sauerkraut which has a softer texture than normal.

Spoilage and Defects in the Sauerkraut Process

The majority of spoilage in sauerkraut is due to aerobic soil micro-organisms which break down the protein and produce undesirable flavour and texture changes. The growth of these aerobes can easily be inhibited by a normal fermentation.

Soft kraut can result from many conditions such as large amounts of air, poor salting procedure and varying temperatures. Whenever the normal sequence of bacterial growth is altered or disturbed, it usually results in a soft product. It is the lactobacilli, which seem to have a greater ability than the cocci to break down cabbage tissues, which are responsible for the softening. High temperatures and a reduced salt content favour the growth of lactobacilli, which are sensitive to higher concentrations of salt. The usual concentration of salt used in sauerkraut production slightly inhibits the lactobacilli, but has no effect on the cocci. If the salt content is too low initially, the lactobacilli grow too rapidly at the beginning and upset the normal sequence of fermentation.

Another problem encountered is the production of dark coloured sauerkraut. This is caused by spoilage organisms during the fermentation process. Several conditions favour the growth of spoilage organisms. For example, an uneven distribution of salt tends to inhibit the desirable organisms while at the same time allowing the undesirable salt tolerant organisms to flourish. An insufficient level of juice to cover the kraut during the fermentation allows undesirable aerobic bacteria and yeasts to grow on the surface of the kraut, causing off flavours and discoloration. If the fermentation temperature is too high, this also encourages the growth of undesirable microflora, which results in a darkened colour.

Pink kraut is a spoilage problem. It is caused by a group of yeasts which produce an intense red pigment in the juice and on the surface of the cabbage. It is caused by an uneven distribution of or an excessive concentration of salt, both of which allow the yeast to multiply. If conditions are optimal for normal fermentation, these spoilage yeasts are suppressed.

Brine Salted Fermented Vegetables

Brine is used for vegetables which inherently contain less moisture. A brine solution is prepared by dissolving salt in water (a 15 to 20% salt solution). Fermentation takes place well in a brine of about 20 salometer. As a general guide, a fresh egg floats in a 10% brine solution (Kordylas, 1990). Properly brined vegetables will keep well in vinegar for a long time. The duration of brining is important for the overall keeping qualities. The vegetable is immersed in the brine and allowed to ferment. The strong brine solution draws sugar and water out of the vegetable, which decreases the salt concentration. It is crucial that the salt concentration does not fall below 12%, otherwise conditions do not allow for fermentation. To achieve this, extra salt is added periodically to the brine mixture.

Once the vegetables have been brined and the container sealed, there is a rapid development of micro-organisms in the brine. The natural controls which affect the microbial populations of the fermenting vegetables include the concentration of salt and temperature of the brine, the availability of fermentable materials and the numbers and types of micro-organisms present at the start of fermentation. The rapidity of the fermentation is correlated with the concentration of salt in the brine and its temperature.

Most vegetables can be fermented at 12.5o to 20o salometer salt. If so, the microbial sequence of lactic acid bacteria generally follows the classical sauerkraut fermentation described by Pederson (1979). At higher salt levels of up to about 40o salometer, the sequence is skewed towards the development of a homofermentation, dominated by *Lactobacillus plantarum*. At the highest concentrations of salt (about 60o salometer) the lactic fermentation ceases to function and if any acid is detected during brine storage it is acetic acid, presumably produced by acid-forming yeasts which are still active at this concentration of salt (Vaughn, 1985).

Brine Salted Fermentation of Vegetables (Pickles)

Pickled cucumbers are another fermented product that has been studied in detail and the process is known. The fermentation process is very similar to the sauerkraut process, only brine is used instead of dry salt. The washed cucumbers are placed in large tanks and salt brine (15 to 20%) is added. The cucumbers are submerged in the brine, ensuring that none float on the surface

- this is essential to prevent spoilage. The strong brine draws the sugar and water out of the cucumbers, which simultaneously reduces the salinity of the solution. In order to maintain a salt solution so that fermentation can take place, more salt has to be added to the brine solution. If the concentration of salt falls below 12%, it will result in spoilage of the pickles through putrefaction and softening.

A few days after the cucumbers have been placed in the brine, the fermentation process begins. The process generates heat which causes the brine to boil rapidly. Acids are also produced as a result of the fermentation. During fermentation, visible changes take place which are important in judging the progress of the process. The colour of the cucumber surface changes from bright green to a dark olive green as acids interact with the chlorophyll. The interior of the cucumber changes from white to a waxy translucent shade as air is forced out of the cells. The specific gravity of the cucumbers also increases as a result of the gradual absorption of salt and they begin to sink in the brine rather than floating on the surface.

Microbes involved in the fermentation process

As with the sauerkraut process, the gram positive coccus - *Leuconostoc mesenteroides* predominates in the first stages of pickle fermentation. This species is more resistant to temperature changes and tolerates higher salt concentration than the subsequent species. As fermentation proceeds and the acidity increases, lactobacilli start to take over from the cocci. The active stage of fermentation continues for between 10 to 30 days, depending upon the temperature of the fermentation. The optimum temperature for *L. Cucumeris* is 29 to 32ºC. During the fermentative period, the acidity increases to about 2% and the strong acid producing types of bacteria reach their maximum growth. If sugar or acetic acid is added to the fermenting mixture during this time it increases the production of acid.

Problems in Pickles

The production of excessive amounts of acid during the fermentation, results in shrivelling of the pickles, possibly due to over-activity of the *L. mesenteroides* species. If the brine is stirred, it may introduce air, which makes conditions more favourable for the growth of spoilage bacteria. In general, if the pickles are well covered with brine, the salt concentration is maintained and the temperature is at an optimum, it should be quite simple to produce good quality pickles.

Non-Salted, Lactic Acid Fermented Vegetables

Some vegetables are fermented by lactic acid bacteria, without the prior addition of salt or brine. Examples of non-salted products include *gundruk* (consumed in Nepal), *sinki* and other wilted fermented leaves. The detoxification of cassava through fermentation includes an acid fermentation, during which time the cyanogenic glycosides are hydrolysed to liberate the toxic cyanide gas.

The fermentation process relies on the rapid colonisation of the food by lactic acid producing bacteria, which lower the pH and make the environment unsuitable for the growth of spoilage organisms. Oxygen is also excluded as the *Lactobacilli* favour an anaerobic atmosphere. Restriction of oxygen ensures that yeasts do not grow.

For the production of *sinki*, fresh radish roots are harvested, washed and wilted by sun-drying for one to two days. They are then shredded, re-washed and packed tightly into an earthenware or glass jar, which is sealed and left to ferment. The optimum fermentation time is twelve days at 30ºC. *Sinki* fermentation is initiated by *L. fermentum* and *L. brevis*, followed by *L. plantarum*. During fermentation the pH drops from 6.7 to 3.3. After fermentation, the radish substrate is sun-dried to a moisture level of about 21%. For consumption, *sinki* is rinsed in water for two minutes, squeezed to remove the excess water and fried with salt, tomato, onion and green chilli. The fried mixture is then boiled in rice water and served hot as soup along with the main meal (Steinkraus, 1996).

Pit fermentations

South Pacific pit fermentations are an ancient method of preserving starchy vegetables without the addition of salt. The raw materials undergo an acid fermentation within the pit, to produce a paste with good keeping qualities. Pit fermentations are also used in other parts of the world – for example in Ethiopia, where the false banana (*Ensete ventricosum*) is fermented in a pit to produce a pulp known as *kocho*. Foods preserved in pits can last for years without deterioration, therefore pits provide a good, reliable cheap means of storage.

Root crops and bananas are peeled before being placed in the pit, while breadfruit are scraped and pierced. Food is left to ferment for three to six weeks, after which time it becomes soft, has a strong odour and a paste-like consistency. During fermentation, carbon dioxide builds up in the pit, creating an anaerobic atmosphere. As a result of bacterial activity, the temperature rises much higher than the ambient temperature.

The pH of the fruit within the pit decreases from 6.7 to 3.7 within about four weeks. Inoculation of the fruit in the pit with lactic acid bacteria greatly speeds up the process. The fermented paste can be left in the pit and removed as required. Usually, it is removed and replaced with a second batch of fresh food to ferment.

The fermented food is washed and fibrous material removed. It is then dried in the sun for several hours to remove the volatile odours, and pounded into a paste. Grated coconut or coconut cream and sugar may be added and the mixture is wrapped in banana leaves and either baked or boiled (Steinkraus, 1996).

Principles of Acetic Acid Fermentation

The main desirable fermentation carried out by acetic acid bacteria is the production of vinegar. Vinegar, literally translated as sour wine, is one of the oldest products of fermentation used by man. It can be made from almost any fermentable carbohydrate source, for example fruits, vegetables, syrups and wine. Whatever the raw material used, the fermentation process follows a definite sequence.

The basic requirement for vinegar production is a raw material that will undergo an alcoholic fermentation. Apples, pears, grapes, honey, syrups, cereals, hydrolysed starches, beer and wine are all ideal substrates for the production of vinegar. The best raw materials are cider and wine, which are widely used in Europe and the United States. To produce a high quality product it is essential that the raw material is mature, clean and in good condition.

Microbes Involved in the Vinegar Process

The production of vinegar depends on a mixed fermentation, which involves both yeasts and bacteria. The fermentation is usually initiated by yeasts which break down glucose into ethyl alcohol with the liberation of carbon dioxide gas. Following on from the yeasts, *acetobacter* oxidise the alcohol to acetic acid and water.

Yeast reaction

$$C_6H_{12}O_6 \rightarrow 2C_2H_5OH + 2CO_2$$

Glu cose yeast ethyl alcohol + Carbondioxide

Bacterial reaction

$$C_2H_5OH + O_2 \rightarrow CH_3COOH + H_2O$$

Alcohol aceticacid water

The yeasts and bacteria exist together in a form known as commensalism. The acetobacter are dependent upon the yeasts to produce an easily oxidisable substance (ethyl alcohol). It is not possible to produce vinegar by the action of one type of micro-organism alone.

For a good fermentation, it is essential to have an alcohol concentration of 10 to 13%. If the alcohol content is much higher, the alcohol is incompletely oxidised to acetic acid. If it is lower than 13%, there is a loss of vinegar because the esters and acetic acid are oxidised. In addition to acetic acid, other organic acids are formed during the fermentation which become esterified and contribute to the characteristic odour, flavour and colour of the vinegar.

Acetaldehyde is an intermediate product in the transformation of the reducing sugar in fruit juice to acetic acid or vinegar. Oxygen is required for the conversion of acetaldehyde to acetic acid. In general, the yield of acetic acid from glucose is approximately 60%. That is three parts of glucose yield two parts acetic acid.

Micro-organisms involved in the fermentation of vinegar

The organisms involved in vinegar production usually grow at the top of the substrate, forming a jelly like mass. This mass is known as 'mother of vinegar'. The mother is composed of both *acetobacter* and yeasts, which work together. The principal bacteria are *Acetobacter acetic A. Xylinum* and *A. Ascendens*. The main yeasts are *Saccharomyces ellipsoideus* and S *cerevisiae*. It is important to maintain an acidic environment to suppress the growth of undesirable organisms and to encourage the presence of desirable acetic acid producing bacteria. It is common practice to add 10 to 25% by volume of strong vinegar to the alcoholic substrate in order to attain a desirable fermentation.

The alcoholic fermentation of sugars should be completed before the solution is acidified because any remaining sugar will not be converted to alcohol after the acetic acid is added. Incomplete fermentation of the juice results in a "weak" product. The acetic acid strength of good vinegar should be approximately 6%.

Fermentation Methods

Small Scale Production

Vinegar can be made at home at the small scale by introducing oxygen into barrels of wine or cider and allowing fermentation to occur spontaneously. This process is not very rigorously controlled and often results in a poor quality product.

The Orleans process

The Orleans process is one of the oldest and well known methods for the production of vinegar. It is a slow, continuous process, which originated in France. High grade vinegar is used as a starter culture, to which wine is added at weekly intervals. The vinegar is fermented in large (200 litre) capacity barrels. Approximately 65 to 70 litres of high grade vinegar is added to the barrel along with 15 litres of wine. After one week, a further 10 to 15 litres of wine are added and this is repeated at weekly intervals. After about four weeks, vinegar can be withdrawn from the barrel (10 to 15 litres per week) as more wine is added to replace the vinegar.

One of the problems encountered with this method is that of how to add more liquid to the barrel without disturbing the floating bacterial mat. This can be overcome by using a glass tube which reaches to the bottom of the barrel. Additional liquid is poured in through the tube and therefore does not disturb the bacteria. Wood shavings are sometimes added to the fermenting barrel to help support the bacterial mat.

Quick vinegar method

Because the Orleans process is slow, other methods have been adapted to try and speed up the process. The German method is one such method. It uses a generator, which is an upright tank filled with beechwood shavings and fitted with devices which allow the alcoholic solution to trickle down through the shavings in which the acetic acid bacteria are living.

The tank is not allowed to fill as that would exclude oxygen which is necessary for the fermentation. Near the bottom of the generator are holes which allow air to be drawn in. the air rises through the generator and is used by the acetic acid bacteria to oxidise the alcohol. This oxidisation also releases considerable amounts of heat which must be controlled to avoid causing damage to the bacteria.

Problems in vinegar production

Many of the problems of vinegar production are concerned with the presence of nematodes, mites, flies and other insects. These pests can be controlled by adherence to good hygiene and pasteurisation of the vinegar. Problems associated with the fermentative process include the presence of a whitish film on the surface of the vinegar.

This is sometimes called *Mycoderma vini* and is composed of yeast-like organisms, which grow aerobically and oxidise the carbon containing compounds to carbon dioxide and water. They also alter the flavour and alcohol content of the vinegar. This problem can however be controlled by adding one part vinegar to three parts of the alcoholic solution or by storing the alcoholic liquid in filled closed containers.

PRODUCTS OF YEAST FERMENTATATION

The major products of yeast fermentation are alcoholic drinks and bread. With respect to fruits and vegetables, the most important products are fermented fruit juices and fermented plant saps. Virtually any fruit or sugary plant sap can be processed into an alcoholic beverage. The process is well known being essentially an alcoholic fermentation of sugars to yield alcohol and carbon dioxide. It should be noted that alcohol production requires special licences or is prohibited in many countries.

Fermented Fruit Juices

There are many fermented drinks made from fruit in Africa, Asia and Latin America. These include drinks made from bananas, grapes and other fruit. Grape wine is perhaps the most economically important fruit juice alcohol. It is of major economic importance in Chile, Argentina, South Africa, Georgia, Morocco and Algeria. Because of the commercialisation of the product for industry, the process has received most research attention and is documented in detail. Banana beer is probably the most wide spread alcoholic fruit drink in Africa and is of cultural importance in certain areas. Alcoholic fruit drinks are made from many other fruits including dates in North Africa, pineapples in Latin America and jack fruits in Asia.

Red Grape Wine

Location of Production

Red grape wines are made in many African, Asian and Latin American countries including Algeria, Morocco and South Africa.

Product Description

Red grape wine is an alcoholic fruit drink of between 10 and 14% alcoholic strength. The colour ranges from a light red to a deep dark red. It is made from the fruit of the grape plant (*Vitis vinifera*). There are many varieties of grape used including Cabernet Sauvignon, Grenache, Nebbiolo, Pinot Noir, and Torrontes (Ranken, Kill and Baker, 1997). The skins of the grape are allowed to be fermented in red wine production, to allow for the extraction of colour and tannins, which contribute to the flavour. The grapes contribute trace elements of many volatile substances, which give the final product the distinctive fruity character. In addition, they contribute non-volatile compounds (tartaric and malic acids) which impact on flavour and tannins, which give bitterness and astringency.

Raw Material Preparation

Ripe and undamaged grapes should be used. Red grapes are crushed to yield the juice plus skins, which is known as must.

Processing

The crushed grapes are transferred to fermentation vessels. The ethanol formed during this fermentation assists with the extraction of pigments from the skins. This takes between 24 hours and three weeks depending on the colour of the final product required.

The skins are then removed and the partially fermented wine is transferred to a separate tank to complete the fermentation. The fermentation can be from naturally occurring yeasts on the skin of the grape or using a starter culture of *Saccharomyces cerevisiae* – in which case the juice is inoculated with populations of yeast. This approach produces a wine of generally expected taste and quality. If the fermentation is allowed to proceed naturally, utilising the yeasts present on the surface of the fruits, the end result is less controllable, but produces wines with a range of flavour characteristics (Fleet, 1998), (Rhodes and Fletcher,1966), (Colquichagua, 1994).

Traditionally, fermentation was carried out in large wooden barrels or concrete tanks. Modern wineries now use stainless steel tanks as these are more hygienic and provide better temperature control. Fermentation stops naturally when all the fermentable sugars have been converted to alcohol or when the alcoholic strength reaches the limit of tolerance of the strain of yeast involved.

Fermentation can be stopped artificially by adding alcohol, by sterile filtration or centrifugation (Ranken, Kill and Baker, 1997). Some wines can be drunk immediately. However most wines develop distinctive favours and aromas by ageing in wooden casks.

White Grape Wine

Location of Production

White grape wines are made in many African, Asian and Latin American countries including Algeria, Morocco and South Africa.

Product Description

White grape wine is an alcoholic fruit drink of between 10 and 14% alcoholic strength. It is prepared from the fruit of the grape plant (*Vitis vinifera*), and is pale yellow in color. There are many varieties used including Airen, Chardonnay, Palomino, Sauvignon Blanc and Ugni Blanc (Ranken, Kill and Baker, 1997). The main difference between red and white wines is the early removal of grape skins in white wine production. The distinctive flavour of grape wine originates from the grapes as raw material and subsequent processing operations. The grapes contribute trace elements of many volatile substances (mainly terpenes) which give the final product the distinctive fruity character.

Preparation of Raw Materials

Ripe and undamaged grapes should be used. The grapes are crushed to yield the juice and the skins are removed and separated out. Sometimes the juice is clarified by allowing it to stand for 24 to 48 hours at 5 to 10° C, by filtering or centrifugation. Pectolytic enzymes may be added to accelerate the breakdown of cell wall tissue and to improve the clarity of juice. Excessive clarification removes many of the natural yeasts and flora. This is beneficial if a tightly controlled induced fermentation is desired, but less so if the fermentation is a natural one. Long periods of settling out however, encourage the growth of natural flora, which can contribute to the fermentation.

Processing

The clarified juice is transferred to a fermentation tank where fermentation either begins spontaneously or is induced by the addition of a starter culture. Traditionally, fermentation was carried out in large wooden barrels or concrete tanks. Modern wineries now use stainless steel tanks as these are more hygienic and provide better temperature control. White wines are fermented at 10 to 18º C for about seven to fourteen days. The low temperature and slow fermentation favours the retention of volatile compounds (Fleet, 1998).

The fermentation can be from naturally occurring yeasts on the skin of the grape or using a starter culture of *Saccharomyces cerevisiae*. This approach produces a wine of generally expected taste and quality. If the fermentation is allowed to proceed naturally, utilising the yeasts present on the surface of the fruits, the end result is less controllable, but produces wines having a range of flavour characteristics. It is likely that natural fermentations are practised widely around the world, especially for home production of wine.

During storage, wines are prone to non-desirable microbial changes. Yeasts, lactic acid bacteria, acetic acid bacteria and fungi can all spoil or taint wines after the fermentation process is completed.

Packaging and Storage

Traditionally wine was delivered to the point of sale in casks. The product is traditionally packaged in glass bottles with corks, made from the bark of the cork oak (*Quercus suber*). The bottles should be kept out of direct sunlight. During storage, wines are prone to non-desirable microbial changes. Yeasts, lactic acid bacteria, acetic acid bacteria and fungi can all spoil or taint wines after the fermentation process is completed.

Banana Beer

Location of Production

Banana beer is made from bananas, mixed with a cereal flour (often sorghum flour) and fermented to an orange, alcoholic beverage. It is sweet and slightly hazy with a shelf-life of several days under correct storage conditions. There are many variations in how the beer is made. For instance *Urwaga* banana beer in Kenya is made from bananas and sorghum or millet and *Lubisi* is made from bananas and sorghum.

Preparation of Raw Materials

Ripe bananas (*Musa* spp.) are selected. The bananas should be peeled. If the peels cannot be removed by hand then the bananas are not sufficiently ripe.

Processing

The first step of the process is the extraction of banana juice. Extraction of a high yield of banana juice without excessive browning or contamination by spoilage micro-organisms and proper filtration to produce a clear product is of great importance. Grass is used as an aid in obtaining clarified juice.

One volume of water is added to every three volumes of banana juice. This makes the total soluble solids low enough for the yeast to act. Cereals are ground and roasted and added to improve the colour and flavour of the final product. The mixture is placed in a container, which is covered in polythene to ferment for 18 to 24 hours.

The raw materials are not sterilised by boiling and therefore provide an excellent substrate for microbial growth. It is essential that proper hygienic procedures are followed and that all equipment is thoroughly sterilised to prevent contaminating bacteria from competing with the yeast and producing acid instead of alcohol. This can be done by cleaning with boiling water or with chlorine solution. Care is necessary to wash the equipment free of residual chlorine as this would interfere with the actions of the yeast. Strict personal hygiene is also essential (Fellows, 1997).

For many traditional fermented products, the micro-organisms responsible for the fermentation are unknown to scientists. However there has been research to identify the micro-organisms involved in banana beer production. The main micro-organism involved, is *Saccharomyces cerevisiae* which is the same organism involved in the production of grape wine. However many other micro-organisms associated with the fermentation have been identified. These varied according to the region of production (Davies, 1994). After fermentation the product is filtered through cotton cloth.

Packaging and Storage

Packaging is usually only required to keep the product for its relatively short shelf-life. Clean glass or plastic bottles are used. The product is kept in a cool place away from direct sunlight.

Cashew wine

Location of Production

Cashew wine is made in many countries in Asia and Latin America.

Product Description

Cashew wine is a light yellow alcoholic drink prepared from the fruit of the cashew tree (*Ancardium occidentale*). It contains an alcohol content of between 6 and 12% alcohol.

Preparation of Raw Materials

In gathering the fruits and transporting them to the workshop, the prime purpose should be to have the fruit arrive in the very best condition possible. Cashew apples are sorted and only mature undamaged cashew apples should be selected. These should be washed in clean water.

Processing

The cashew apples are cut into slices to ensure a rapid rate of juice extraction when crushed in a juice press. The fruit juice is sterilised in stainless steel pans at a temperature of 85oC in order to eliminate wild yeast. The juice is filtered and treated either sodium or potassium metabisulphite to destroy or inhibit the growth of any undesirable types of micro-organisms - acetic acid bacteria, wild yeasts and moulds. Wine yeast (*Saccharomyees cerevisiae - var ellipsoideus*) are added. Once the yeast is added, the contents are stirred well and allowed to ferment for about two weeks. The wine is separated from the sediment. It is clarified by using fining agents such as gelatin, pectin or casein which are mixed with the wine. Filtration is carried out with filter-aids such as fullers earth. The filtered wine is transferred to wooden vats.

The wine is then pasteurised at 50o - 60oC. Temperature should be controlled, so as not to heat it to about 70oC, since its alcohol content would vaporise at a temperature of 75o-78oC. It is then stored in wooden vats and subjected to ageing. At least six months should be allowed for ageing. If necessary, wine is again clarified prior to bottling. During ageing, and subsequent maturing in bottles many reactions, including oxidation, occur with the formation of traces of esters and aldehydes., which together with the tannin and acids already present enhance the taste, aroma and preservative properties of the wine (Wimalsiri, Sinnatamby, Samaranayake and Samarsinghe, 1971).

Packaging and Storage

The product is packaged in glass bottles with corks. The bottles should be kept out of direct sunlight.

Tepache

Tepache is a light, refreshing beverage prepared and consumed throughout Mexico. In the past, *tepache* was prepared from maize, but nowadays various fruits such as pineapple, apple and orange are used. The pulp and juice of the fruit are allowed to ferment for one or two days in water with some added brown sugar. The mixture is contained in a lidless wooden barrel called a *'tepachera'*, which is covered with cheese cloth. After a day or two, the *tepache* is a sweet and refreshing beverage. If fermentation is allowed to proceed longer, it turns into an alcoholic beverage and later into vinegar. The microorganisms associated with the product include *Bacillus subtilis, B. graveolus* and the yeasts, *Torulopsis insconspicna, Saccharomyces cerevisiae* and *Candida queretana* (Aidoo, 1986).

Colonche

Colonche is a sweet, fizzy beverage produced in Mexico by fermenting the juice of the fruits of the prickly pear cacti - mainly *Opuntia* species. The procedure for preparing *colonche* is essentially the same as has been followed for centuries. The cactus fruits are peeled and crushed to obtain the juice, which is boiled for 2-3 hours. After cooling, the juice is allowed to ferment for a few days. Sometimes old *colonche* or *tibicos* may be added as a starter. *Tibicos* are gelatinous masses of yeasts and bacteria, grown in water with brown sugar. They are also used in the preparation of *tepache*.

Fortified Grape Wines

Fortified wines are made in the Republic of South Africa and North Africa. Fortified wines are made by adding spirits to wines, either during or after fermentation, with the result that the alcohol content of the wines is raised to around 20 percent, i.e. approximately double that of table wines (Rose, 1961).

Date Wine

Date wines are popular in Sudan and North Africa. They are made using a variety of methodologies. *Dakhai* is produced by placing dates in a clean earthenware pot. For every one volume of dates between two and four volumes of boiling water are added. This is allowed to cool and is then sealed for three days. More warm water is then added and the container sealed again for seven to ten days. Many variations of date wine exist: *El madfuna* is produced by burying the earthenware pots underground. *Benti merse* is produced from a mixture of sorghum and dates. *Nebit* is produced from date syrup (Dirar,1992).

Sparkling Grape Wine

Sparkling grape wines are made in the Republic of South Africa. Sparkling wines can be made in one of three ways. The cheapest method is to carbonate wines under pressure. Unfortunately, the sparkle of these wines quickly disappears, andthe product is considered inferior to the sparkling wines produced by the traditional method of secondary fermentation. This involves adding a special strain of wine yeast (*S. cerevisiae* var. *ellipsoideus*) - a champagne yeast - to wine that has been artificially sweetened. Carbon dioxide produced by fermentation of the added sugar gives the wine its sparkle. In the original champagne method, which is still widely used today, this secondary fermentation is carried out in strong bottles, capable of withstanding pressure but early in the nineteenth century a method of fermenting the wine in closed tanks was devised, this being considerably cheaper than using bottles (Rose, 1961).

Jack-Fruit Wine

Jack-fruit wine is an alcoholic beverage made by ethnic groups in the eastern hilly areas of India. As its name suggests, it is produced from the pulp of jack-fruit (*Artocarpus heterophyllus*). Ripe fruit is peeled and the skin discarded. The seeds are removed and the pulp soaked in water. Using bamboo baskets, the pulp is ground to extract the juice, which is collected in earthenware pots. A little water is added to the pots along with fermented wine inoculum from a previous fermentation. The pots are covered with banana leaves and allowed to ferment at 18 to 30ºC for about one week. The liquid is then decanted and drunk. During fermentation, the pH of the wine reaches a value of 3.5 to 3.8, suggesting that an acidic fermentation takes place at the same time as the alcoholic fermentation. Final alcohol content is about 7 to 8% within a fortnight (Steinkraus, 1996).

Fermented Plant Saps

Virtually any sugary plant sap can be processed into an alcoholic beverage. The process is well known being essentially an alcoholic fermentation of sugars to yield alcohol and carbon dioxide. Many alcoholic drinks are made from the juices of plants including coconut palm, oil palm, wild date palm, nipa palm, raphia palm and kithul palm.

Palm Wine

Location of Production

Palm 'wine' is an important alcoholic beverage in West Africa where it is consumed by more than 10 million people.

Product Description

Palm wine can be consumed in a variety of flavours varying from sweet unfermented to sour fermented and vinegary alcoholic drinks. There are many variations and names including *emu* and *ogogoro* in Nigeria and *nsafufuo* in Ghana. It is produced from sugary palm saps. The most frequently tapped palms are raphia palms (*Raphia hookeri* or *R. vinifera*) and the oil palm (*Elaeis guineense*). Palm wine has been found to be nutritious. The fermentation process increases the levels of thiamin, riboflavin, pyridoxin and vitamin B12. Like many African alcoholic beverages, palm wine has a very short shelf-life. The product is not preserved for more than one day. After this time accumulation of an excessive amount of acetic acid makes it unacceptable to consumers. The bark of a tree (*Saccoglottis gabonensis*) may be added as a preservative. The alkaloid and phenolic compounds which are extracted into the wine have antimicrobial effect (Odunfa, 1985) .

Preparation of Raw Materials

Sap is collected by tapping the palm. Tapping is achieved by making an incision between the kernels and a gourd is tied around to collect the sap which is collected a day or two later. The fresh palm juice is a sweet, clear, colourless juice containing 10-12 percent sugar and is neutral. The quality of the final wines is determined mostly by the conditions used in the collection of the sap. Often the collecting gourd is not washed between collections and residual yeasts in the gourd quickly begin the fermentation.

Processing

The sap is not heated and the wine is an excellent substrate for microbial growth. It is therefore essential that proper hygienic collection procedures are followed to prevent contaminating bacteria from competing with the yeast and producing acid instead of alcohol (Fellows, 1997).

Fermentation starts soon after the sap is collected and within an hour or two, the sap becomes reasonably high in alcohol (up to 4%). If allowed to continue to ferment for more than a day, the sap begins turning into vinegar, although the vinegary flavour is preferred by some. Organisms responsible include *S. cerevisiae,* and *Schizosaccharomyces pombe,* and the bacteria *Lactobacillus plantarum* and *L. mesenteroides*. There are reports that the yeasts and bacteria originate from the gourd, palm tree, and tapping implements. However the high sugar content of the juice would seem to selectively favour the growth of yeasts which might originate from the air. This is supported by the fact that fermentation also takes place in plastic containers. Within 24 hours the initial pH is reduced from 7.4-6.8 to 5.5 and the alcohol content ranges from 1.5 to 2.1 percent. Within 72 hours the alcohol levels increase from 4.5 to 5.2 percent and the pH is 4.0. Organic acids present are lactic acid, acetic acid and tartaric acid (Odunfa, 1985).

The main control points are extraction of a high yield of palm sap without excessive contamination by spoilage micro-organisms, and proper storage to allow natural fermentation to take place.

Packaging and Storage

Packaging is usually only required to keep the product for its relatively short shelf-life. Clean glass or plastic bottles should be used. The product should be kept in a cool place away from direct sunlight.

Toddy

Location of Production

Throughout Asia, particularly India and Sri Lanka.

Product Description

Toddy is an alcoholic drink made by the fermentation of the sap from a coconut palm. It is white and sweet with a characteristic flavour. It is between 4 and 6% alcohol and has a shelf life of about 24 hours.

Preparation of Raw Materials

The sap is collected by slicing off the tip of an unopened flower. The sap oozes out and can be collected in a small pot tied underneath the flower.

Processing

The fermentation starts as soon as the sap collects in the pots on the palms, particularly if a small amount of toddy is left in the pots. The toddy is fully fermented in six to eight hours. The product is usually sold immediately due to its short shelf-life (Fellows, 1997) .

Packaging and Storage

Packaging is usually only required to keep the product for its relatively short shelf-life. This is usually clean glass or plastic bottles. The product should be kept in a cool place away from direct sunlight.

Pulque

Location of Production

Pulque is the national drink in Mexico, where, it is claimed, it originated with the early Aztecs. *Pulque* is a traditional beverage that now forms the basis of a national industry, together with the spirits mezcal and tequila that are obtained from it. Pulque plays an important role in the nutrition of low income people in Mexico with B vitamins being present in nutritionally important levels.

Product Description

Pulque is a milky, slightly foamy, acidic and somewhat viscous beverage. It is obtained by fermentation of aguamiel, which is the name given to the juices of various cacti, notably *Agave atrovirens* and *A. americana* which are often called the "Century plant" in English. The alcohol content on pulque varies between six and seven percent. The beverage obtained upon distilling pulque is called "Mezcal", and if manufactured in the Tequila region from a numbered distillery, it is referred to as "Tequila". The drink is often considered an aphrodisiac. The name Ticyaol is given to a good strain that makes one particularly virile. Pulque is frequently the potion of choice used by women during menstruation.

Preparation of Raw Materials

The juices are extracted from the plants when they are eight to ten years old and fermentation takes place spontaneously, although occasionally the juices are inoculated with a starter from previous fermentations.

Processing

The juice is allowed to ferment naturally through a mixed fermentation although yeast (*Saccharomyces carbajali)* is the main actor. *Lacto-bacillus plantarum* produces lactic acid and the viscosity of pulque is caused by the activity of two species of *Leuconostoc* which produce dextrans (Wood and Hodge).

During fermentation of the juices of the plant, the soluble solids are reduced from between 25-30% to 6%; the pH falls from 7.4 to between 3.5 and 4.0; the sucrose content falls from 15% to 1% and vitamin levels are increased. For instance the vitamin content (milligrams of vitamins per 100g of product) increases from 5 to 29 for thiamine, 54 to 515 for niacin and 18 to 33 for riboflavin (Steinkraus, K.H. (1992) .

Packaging and Storage

Packaging is only required to keep the product for its relatively short shelf-life. Clean glass or plastic bottles should be used. The product should be kept in a cool place away from direct sunlight.

Ulanzi (Bamboo Wine)

Location of Production

East and Southern Africa.

Product Description

Ulanzi is a fermented bamboo sap obtained by tapping young bamboo shoots during the rainy season. It is a clear, whitish drink with a sweet and alcoholic flavour.

Preparation of Raw Materials

The bamboo shoots should be young in order to obtain a high yield of sap. The growing tip is removed and a container fixed in place to collect the sap. The container should be clean in order to prevent contamination of the fresh sap.

Processing

The raw material is an excellent substrate for microbial growth and fermentation begins immediately after collection. Fermentation takes between five and twelve hours depending on the strength of the final product desired.

Packaging and Storage

Packaging is usually only required to keep the product for its relatively short shelf life.

Basi *(Sugar Cane Wine)*

Basi is a sugar cane wine made in the Philipppines by fermenting boiled, freshly extracted, sugar cane juice. A dried powdered starter is used to initiate the fermentation. The mixture is allowed to ferment for up to three months, and to age for up to one year. The final product is light brown in colour and has a sweet and a sour flavour. A similar product called *shoto sake* is made in Japan (Steinkraus, 1996).

Muratina

Muratina is an alcoholic drink made from sugar cane and *muratina* fruit in Kenya. The fruit is cut in half, sun dried and boiled in water. The water is removed and the fruit is again sun dried.

The fruit is added to a small amount of sugar cane juice and incubated in a warm place for 24 hours, after which it is removed and sun dried. The dried fruit is then added to a barrel of sugar cane juice which is allowed to ferment for between one and four days. The final product has a sour alcoholic taste.

Yoghurt

Yoghurt or yogurt, or less commonly yoghourt, joghurt or yogourt, is a dairy product produced by bacterial fermentation of milk. Any sort of milk may be used to make yoghurt, but modern production is dominated by cow's milk.

It is the fermentation of milk sugar (lactose) into lactic acid that gives yoghurt its gel-like texture and characteristic tang. It is often sold in a fruit, vanilla, or chocolate flavour, but can also be unflavoured.

Fig. Yoghurt

Fig. Yoghurt Sold at the Bulgarija Pavilion of Expo 2005 Aichi Japan

Fig. Yoghurt Sold at the Caucasus Common Pavilion of Expo 2005 Aichi Japan

History

There is evidence of cultured milk products being produced as food for at least 4,500 years, since the 3rd millennium BC. The Bulgars (also Hunno-

Bulgars), a Turkic-speaking people from Aryian-Pamirian origin, migrated into Europe starting from the 2nd century AD, eventually settling on the Balkans by the end of the 7th century AD. The earliest yoghurts were probably spontaneously fermented, perhaps by wild bacteria residing inside goat skin bags used for transportation.

Yoghurt remained primarily a food of India, Central Asia, Western Asia, South Eastern Europe and Central Europe until the 1900s, when a Russian biologist named Ilya Ilyich Mechnikov theorized that heavy consumption of yoghurt was responsible for the unusually long lifespans of Bulgarian peasants.

Believing *Lactobacillus* to be essential for good health, Mechnikov worked to popularize yoghurt as a foodstuff throughout Europe. It fell to a Spanish entrepreneur named Isaac Carasso to industrialise the production of yoghurt. In 1919 he started a commercial yoghurt plant in Barcelona, naming the business Danone after his son — better known in the United States as 'Dannon'.

Yoghurt with added fruit marmalade was invented (and patented) in 1933 in dairy Radlická Mlékárna in Prague. The original intention of this combination was to protect yoghurt better against decay. Yoghurt was first commercially produced and sold in the United States in 1929 by Armenian immigrants, Rose and Sarkis Colombosian, whose family business later became Colombo Yogurt.

Contents

Yoghurt making involves the introduction of specific "friendly" bacteria into milk under controlled (very carefully in industrial settings) temperature and environmental conditions. The bacteria ingest the natural milk sugars and release lactic acid as a waste product; the increased acidity, in turn, causes the milk proteins to tangle into a solid mass, (curd, denature).

The increased acidity (pH=4–5) also prevents the proliferation of other potentially pathogenic bacteria. To be named yoghurt, the product should at least contain the bacteria *Streptococcus salivarius ssp thermophilus* and *Lactobacillus bulgaricus* (official name *Lactobacillus delbrueckii* ssp. *bulgaricus*). Often these are co-cultured with other lactic acid bacteria for either taste or health effects (probiotics). These include *L. acidophilus, Lactobacillus casei and* Bifidobacterium *species.*

In most countries a product may only be called yoghurt if there are live bacteria present in the final product. Pasteurized products (which have no living bacteria) are named fermented milk (drink). In the US non-pasteurized yoghurt is sold as containing "live active culture" (or just as "live"), which some believe to be nutritionally superior. In Spain, the yoghurt producers were divided among those who wanted to reserve the name *yogurt* for live yoghurt and those who wanted to include pasteurised yoghurt under that label (mostly the Pascual Hermanos group). Pasteurized yoghurt has a shelf life of months and does not require refrigeration. Both sides submitted scientific studies claiming differences or lack thereof between both varieties. Eventually the Spanish government allowed the label *yogur pasteurizado* instead of the former *postre lácteo* ("dairy dessert").

Because live yoghurt culture contains enzymes that break down lactose, some individuals who are otherwise lactose intolerant find that they can enjoy yoghurt without ill effects. Nutritionally, yoghurt is rich in protein as well as several B vitamins and essential minerals, and it is as low or high in fat as the milk it is made from.

Non-sweetened drinkable yoghurt is typically sold in the West under the name "(cultured) buttermilk." This term is a misnomer, as the drink has little in common with "true" buttermilk and is, in fact, most similar to kefir.

Presentation

Yoghurt is often sold sweetened and flavoured, or with added fruit on the bottom (sometimes referred to as fruit bottom), to offset its natural sourness. If the fruit is already stirred into the yoghurt, it is sometimes referred to as Swiss-style. Most yoghurt in the United States has pectin or gelatin added. Some specialty yoghurts have a layer of fermented fat at the top similar to cream cheese (e.g. brands like Brown Cow Yoghurt). Fruit jam is used instead of raw fruit pieces in the case of fruit yoghurts so that they can be stored for weeks.

Yoghurt Types

Dahi Yoghurt

Dahi yoghurt of the Indian subcontinent is known for its characteristic taste and consistency. The English term for yoghurt in India is curd.

Labneh or Labaneh

Labneh yoghurt of Lebanon is a thickened yogurt that is used to make sandwiches by itself, or with olive oil, cucumber slices, olives, and various green herbs. It is also thickened further and rolled into balls that are preserved in olive oil, and fermented further for a few weeks before it is eaten. It is sometimes used with onions, meat, and nuts as a stuffing for a variety of Lebanese pies or Kebbeh ßÈÉ balls.

Bulgarian Yoghurt

Bulgarian yoghurt is popular for its specific taste, aroma, and quality and is commonly consumed plain. The qualities are specific to the particular culture strains used in Bulgaria, *Lactobacillus bulgaricus* and *Streptococcus thermophilus* bacteria. Bulgarian yoghurt producers are taking steps to legally protect the trademark of Bulgarian yoghurt on the European market and distinguish it from other product types that do not contain live bacteria.

Bulgarian yoghurt is often strained by hanging in a cloth for a few hours to reduce water content. The resulting yoghurt is creamier, richer and milder in taste because of increased fat content. Hanging overnight is sometimes employed to make a concentrated yoghurt similar to cream cheese. Yoghurt is also used for preparation of Bulgarian milk salad. (Commercial versions of strained yoghurt are also made.) A cold soup (called tarator in Bulgaria and cacýk in Turkey) made of yoghurt is popular in Turkey and Bulgaria in the summertime. It is made from Ayran, cucumbers, garlic and ground walnuts.

Yoghurt Drinks

Lassi is a yoghurt-based beverage, originally from India where two basic varieties are known: salty and sweet. Salty lassi is usually flavoured with ground-roasted cumin and chile peppers; the sweet variety with rosewater and/or lemon, mango, or other fruit juice. Another yoghurt-based beverage, a salty drink called Ayran is quite popular in Turkey, Bulgaria and Greece.

It is made by mixing yoghurt with water and adding salt. The same drink is known as *tan* in Armenia. A similar drink, Doogh, is popular in the Middle East between Lebanon and Iran; it differs from ayran by the addition of herbs (usually mint) and being carbonated (usually with seltzer water). In the United States, yoghurt-based beverages are often marketed under names like "Yogurt Smoothie" or "Drinkable Yogurt".

Kefir

Kefir is a fermented milk drink originating in the Caucasus. A related Central Asian-Mongolian drink made from mare's milk is called kumis or, in Mongolia, airag. Some American dairies have offered a drink called "kefir" for many years (though lacking the carbonation and alcohol, and coming in fruit flavours), but began appearing (as of 2002) with names like "drinkable yoghurt" and "yoghurt smoothie".

Home-made yoghurt

Home-made yoghurt is consumed by many people throughout the world, and is the norm in countries where yoghurt has an important place in traditional cuisine, such as Bulgaria, Turkey, and India. Yoghurt can be made at home using a small amount of store-bought plain live active culture yoghurt as the starter culture.

One very simple recipe starts with a litre of low-fat milk, but requires some means to incubate the fermenting yoghurt at a constant 43°C (109°F) for several hours. Yoghurt-making machines are available for this purpose. A run of the mill heating pad found in a pharmacy for muscle aches (set at medium), with a pot of tepid water on top to place the milk in, works fine.

As with all fermentation processes, cleanliness is very important:

- Bring the milk to 85°C (185°F) over a stove and keep it there for two minutes, to kill any undesirable microbes.
- Pour the re-pasteurised milk into a tall, sterile container and allow to cool to 43°C (110°F)
- Mix in 120 ml of the warmed yoghurt and cover tightly.
- After about six hours of incubation at precisely 43°C (110°F); the entire mixture will have become a very plain but edible yoghurt with a loose consistency.
  - The further below 43°C (110°F) the temperature, the longer it will take for the yoghurt to solidify. If a precise means of temperature control is not available, put the culture in a warm place such as on top of a water heater or in a gas oven with just the pilot flame burning, or wrap a small towel around the container. An electric oven with the light on may work nicely, depending on the bulb size. You can tell it is done when it no longer moves if you tilt the jar.

In Japan, *Caspian Sea Yoghurt* is a very popular home-made yoghurt. It is believed to have been introduced into the country by researchers in the form of a sample brought back from Georgia in the Caucasus region in 1986. This Georgian yogurt, called *Matsoni* which is mostly made up of *Lactococcus lactis* subsp. *cremoris* and *Acetobacter orientalis* has a unique viscous, honey-like texture and is milder in taste than many other yoghurts.

Caspian Sea yoghurt is particularly well suited for making at home because it does not require any special equipment and cultures at room temperature (20–30°C) in about 10 to 15 hours, depending on the temperature.

In Japan it is possible to buy a freeze-dried starter culture at big department stores or online, but many people obtain a quantity of the yoghurt from a friend and start making their own yoghurt from that.

- General instructions: sterilise all utensils, containers and lids in boiling water prior to use.
  - From freeze-dried starter: in a container stir starter into about 250 ml of milk and cover with a lid. Incubation time is approximately 12-36 hours from starter. Make the next batch as below (from the actual yoghurt as the starter).
  - From yoghurt: In winter, use about one part yoghurt to four parts milk. In summer use about one part yoghurt to nine parts milk.
- Place the lid gently on top of the container so as to allow some air in, but prevent contamination. Leave in a clean dry place for 10-15 hours or until thick. In summer, this may be less than 10 hours and in winter, longer than 15 hours.
- Some thickening of the yoghurt will also occur in the refrigerator.
- The yoghurt can be stored in the refrigerator for about 1 week or longer.
- To reduce contamination, always make the next batch of yoghurt before using the current batch, and use containers solely for making yoghurt and nothing else.

Vinegar

Vinegar (from Old French *vinaigre*, meaning "sour wine") is a sour-tasting liquid made from the oxidation of ethanol in wine, cider, beer, fermented fruit juice, or nearly any other liquid containing alcohol. It can also be made by certain bacteria operating on sugar-water solutions directly, without intermediary conversion to ethanol. Commercially available vinegar usually has a pH of about 2.4.

Fig. Vinegar is Often Infused with Spices or Herbs—as here, with Oregano

Production

Vinegar production may be started by the addition of mother of vinegar to wine or cider. Vinegar is a dilute form of acetic acid, ranging typically from three to five percent by volume for table vinegar and higher concentrations for pickling. Natural vinegars also contain smaller amounts of tartaric acid, citric acid, and other acids. It has been used since ancient times, and is an important element in Western and European, Asian, and other traditional cuisines of the world.

The oxidation is carried out by acetic acid bacteria, as was shown in 1864 by Louis Pasteur. Modern systems work with vinegar bacteria at the liquid and bringing the air into the vinegar with a venturi pump system or with a turbine. These systems have a production time between 38 hours and three days to get the ready vinegar.

Culinary Uses

Vinegar is commonly used in food preparations, particularly in pickling processes, vinaigrettes, and other salad dressings. It is an ingredient in sauces such as mustard, ketchup, and mayonnaise. It is also often used as a condiment.

Types of Vinegar

White

So-called "white vinegar" (actually clear) can be made by oxidating a distilled alcohol. Alternatively, it may be nothing more than a solution of acetic acid in water. It is used for culinary as well as cleaning purposes.

Malt

Malt vinegar is made by malting barley, causing the starch in the grain to turn to maltose. An ale is then brewed from the maltose and allowed to turn into vinegar, which is then aged. A cheaper alternative, called 'non-brewed condiment', is a solution of 4-8% acetic acid colored with caramel. There is also around 1-3% citric acid present. Australians, British, Americans and Canadians commonly use malt vinegar on fish and chips.

Wine

Wine vinegar is made from red or white wine, and is the most commonly used vinegar in Mediterranean countries, Germany, and other countries. As with wine, there is a considerable range in quality. Better quality wine vinegars are matured in wood for up to two years and exhibit a complex, mellow flavor. Champagne vinegar is made from champagne, and Sherry vinegar, produced in Spain, is made from Sherry. These last two are correspondingly expensive.

Apple Cider

Fig. Shaw's Brand Apple Cider Vinegar

Apple cider vinegar, sometimes known simply as cider vinegar, is made from cider or apple must, and is often sold unfiltered, with a brownish-yellow color. It is currently very popular, partly due to its alleged beneficial health and beauty properties. Some countries, like Canada, prohibit the selling of vinegar over a certain percentage acidity. In terms of cooking, cider vinegar is not good for delicate sauces, but is excellent for use in chutneys and marinades. It can be used to pickle foods, but will darken light fruits and vegetables.

Fruit

Fruit vinegars are made from fruit wines without any additional flavouring. Common flavors of fruit vinegar include black currant, raspberry, and quince. Typically, the flavors of the original fruits remain tasteable in the final vinegar. Most such vinegars are produced in Europe, where there is a growing market for high price vinegars made solely from specific fruits (as opposed to non-fruit vinegars which are infused with fruits or fruit flavors. Vinegars that are not actually made from fruit, but which are flavored with fruit or fruit flavors, are not true vinegars, but are instead classified as flavored vinegars.

Balsamic

Balsamic vinegar is an aromatic, aged type of vinegar manufactured in Modena, Italy. Its flavor is rich, sweet, and complex, with the finest grades being the end product of years of aging in a successive number of casks made of various types of wood (including chestnut, cherry, and juniper). Originally an artisanal product available only to the Italian upper classes, balsamic vinegar became widely known and available around the world in the late 20th century. Now it is very common.

Rice

The Japanese prefer a more delicate rice vinegar and use it for much the same purposes as Europeans, as well as for sushi rice, in which it is an essential ingredient. Rice vinegar is available in white, red and black variants, the last of which is most popular in China. Black rice vinegar may be used as a substitute for balsamic vinegar, though its dark color and the fact that it is aged may be the only similarity between the two products. Some types of rice vinegar are sweetened or otherwise seasoned.

Coconut

Coconut vinegar, made from the sap, or "toddy," of the coconut palm, is used extensively in Southeast Asian cuisine (particularly in the Philippines, a major producer of the product), as well as in some cuisines of India. A cloudy white liquid, it has a particularly sharp, acidic taste with a slightly yeasty note.

Cane

Cane vinegar, made from sugar cane juice, is most popular in the Philippines (where it is called *sukang iloko*), although it is also produced in France and the United States. It ranges from dark yellow to golden brown in color and has a mellow flavor, similar in some respects to rice vinegar, though with a somewhat "fresher" taste. Contrary to expectation, it is not sweeter than other vinegars, containing no residual sugar.

Raisin

Vinegar made from raisins is used in cuisines of the Middle East, and is produced in Turkey. It is cloudy and medium brown in color, with a mild flavor.

Beer

Vinegar made from beer is produced in Germany, Austria, and the

Netherlands. Although its flavor depends on the particular type of beer from which it is made, it is often described as having a malty taste. That produced in Bavaria is a light golden color, with a very sharp and not overly complex flavor.

Honey

Vinegar made from honey is rare, though commercially available honey vinegars are produced in Italy and France.

Chinese Black

Chinese black vinegar is an aged product made from rice, wheat, millet, or sorghum. It has an inky black color and a complex flavor.

Flavored Vinegars

Popular *fruit-flavored vinegars* include those infused with whole raspberries, blueberries, or figs (or else from flavorings derived from these fruits). Some of the more exotic fruit-flavored vinegars include blood orange and pear.

Herb vinegars are flavored with herbs, most commonly Mediterranean herbs such as thyme or oregano. Such vinegars can be prepared at home by adding sprigs of fresh or dried herbs to store-bought vinegar; generally a light- colored, mild tasting vinegar such as that made from white wine is used for this purpose.

An East Asian variety of flavored vinegar known as *sweetened vinegar* is made from rice wine and herbs including ginger, cloves and other spices. It is an integral ingredient in the traditional Chinese postnatal health and celebratory dish of Pork Knuckles and Ginger Stew

Non-Culinary Uses

Cleaning

Vinegar can be a potent, inexpensive and environment-friendly cleaning agent. White vinegar is generally recommended when vinegar is being used as a cleaning fluid.

- A few tablespoons of white vinegar mixed with a few teaspoons of common table salt makes an excellent cleanser for cleaning badly-stained stainless cookware. This vinegar and salt mixture can also remove oxidation from copper-clad cookware and make it shine with practically no rubbing required.
- One cup white vinegar to four cups water (for a stronger solution, one cup white vinegar to one cup water works) makes a fine window-

washing fluid, substituting for Windex. If windows appear streaky after washing with vinegar, add a half-teaspoon of liquid soap to the mix—this removes the waxy, streak-causing residue left over by commercial window cleaners.

- Drains can be cleaned by using a combination of vinegar and baking soda. Pour one-half cup baking soda down the drain, followed by half a cup of white vinegar. Cover the drain while it works, then rinse with several gallons of water.
- Vinegar also works well as a fabric softener; just add half a cup to the rinse cycle.
- Removing odors using commercial cleaners often causes damage to surfaces. Vinegar can act as a very effective odor-remover especially in situations involving sensitive surfaces.
- Weak solutions of vinegar or acetic acid in water are used for douches.
- Vinegar can also be used as a solvent for removing the adhesive residue tapes leave on glass and plastic. It works well for quickly removing an adhesive residue that has been left on for about 1-2 weeks. To remove the adhesive residue using vinegar, you should:
  1. Apply the vinegar to a cloth or paper towel.
  2. Dab the cloth or paper towel in a little bit of vinegar.
  3. Scrub at the surface with the damp side of the cloth or paper towel.
  4. Dry the surface.

Medicine

Vinegar is a folk medicine used in China to prevent the spread of virus such as SARS (Severe Acute Respiratory Syndrome) and other pneumonia outbreaks:

"On February 13, 2003 news of a type of atypical pneumonia that appeared in six cities of south China's Guangdong province has been brought under control, with no cases reported since Monday. According to press conferences held by the Guangdong and Guangzhou governments, local governments at various levels have taken emergency measures to control the prices of isatis root, vinegar and other related anti-virus medicines, which saw soaring prices due to their effectiveness in curing this disease" (source unknown).

The therapeutic use of vinegar is recorded in the second verse of the nursery rhyme "Jack and Jill": "Went to bed and bound his head / With vinegar and brown paper." As with all nursery rhymes there is truth in the story and this one comes from the village of Kilmersdon in Somerset. The vinegar used would likely have been cider vinegar. Apple cider vinegar (ACV) is a much more useful astringent than ice and will reduce inflammation, bruising and swelling in approximately a third of the time that ice will take. Application is directly onto the skin with a flannel, and left on for an hour or so.

Vinegar along with hydrogen peroxide (H2O2) is used in the livestock industry to kill bacteria and viruses before refrigeration storage. A chemical mixture of peracetic acid is formed when acetic acid is mixed with hydrogen peroxide. It is being used in some Asian countries by aerosol sprays for control of pneumonia. A mixture of five-percent acetic acid and three-percent hydrogen peroxide is commonly used. Apple cider vinegar in particular is often touted as a medical aid, from cancer prevention to alleviation of joint pain to weight loss.These claims began in Biblical times; in 1958, Dr. D. C. Jarvis made the remedy popular with a book that sold 500,000 copies.

Claims that cider vinegar can be used as a beauty aid also persist, despite the fact that apple cider vinegar can sometimes be very dangerous to the eyes. The acid will burn and the eyes will become red, but there are no damages to the eyes ever described. If the vinegar contains mother of vinegar the slime bacteria of the mother can cause ophthalmitis.

Many believe that vinegar is also a cure to mild to moderate sunburn when soaked on the area with a towel or in a bath. Cider vinegar is also claimed to be a solution to dandruff, in that the acid in the vinegar kills the fungus *Malassezia furfur* (formerly known as *Pityrosporum ovale*) and restores the chemical balance of the skin. Two tablespoons of vinegar before a meal was found to prevent blood sugar spikes in a study by Carol Johnston, a professor of nutrition at Arizona State University .

Miscellaneous

When vinegar is added to sodium bicarbonate (baking soda), it produces a very fizzy and volatile mixture of carbonic acid decomposing into carbon dioxide and water. It is exemplified as the typical acid-base reaction in school science experiments.

The salt that is formed is sodium acetate:

- Neutralizes lye, a strong base.
- According to the Prophet Mohammed, vinegar is one of the best condiments .
- Lord Byron would consume vast quantities of white vinegar to keep his complexion pale.

- Vinegar can also be used as an organic herbicide.

Beer

Beer is the world's oldest and most popular alcoholic beverage, selling over 133,000 million litres a year. It is produced through the fermentation of starch-based material, commonly barley, though cassava root in Africa, potato in Brazil, and agave in Mexico, among other starch sources, have been used. Only beverages produced by this method are considered to be beer. Neither alcoholic beverages made from the fermentation of sugars derived from nonstarch sources (e.g., grape juice or honey), *nor* beverages which are distilled after fermentation should be classified as such.

Fig. A Selection of Bottled Beers

Because the ingredients and procedures used to make beer can differ, characteristics such as taste and colour may also vary. While local names for beers made with the same methods and ingredients may vary, the similarities of method and ingredients can be detected to form a study of the nature of beer styles.

History

Fig. Egyptian Woman Making Beer (Cairo Museum)

Beer is one of the oldest beverages humans have produced, dating back to at least the 5th millennium BC (prior even to writing), and recorded in the written history of Ancient Egypt and Mesopotamia. As almost any substance containing carbohydrates, namely sugar or starch, can naturally undergo fermentation, it is likely that beer-like beverages were independently invented among various cultures throughout the world.

Beer largely remained a homemaker's activity, made in the home in medieval times. By the 14th and 15th centuries, beermaking was gradually changing from a family-oriented activity to an artisan one, with pubs and monasteries brewing their own beer for mass consumption. Today, the brewing industry is a huge global business, consisting of several multinational companies, and many thousands of smaller producers ranging from brewpubs to regional breweries.

The Brewing Process

Though the process of brewing beer is complex and varies considerably, the basic stages that are consistent are outlined below. There may be additional filtration steps between stages.

The Brewer, designed and engraved, in the Sixteenth Century, by J. Amman.

1. Mashing: The first phase of brewing, in which the malted grains are crushed and soaked in warm water in order to create a malt extract. The mash is held at constant temperature long enough for enzymes to convert starches into fermentable sugars.
2. Sparging: Water is filtered through the mash to dissolve the sugars. The darker, sugar-heavy liquid is called the wort.
3. Boiling: The wort is boiled along with any remaining ingredients (excluding yeast), to remove excess water and kill any microorganisms. The hops (whole, pelleted, or extract) are added at some stage during the boil.
4. Fermentation: The yeast is added (or *"pitched"*) and the beer is left to ferment. After primary fermentation, the beer may be allowed a second

fermentation, which allows further settling of yeast and other particulate matter *"trub"* which may have been introduced earlier in the process. Some brewers may skip the secondary fermentation and simply filter off the yeast.

5. Packaging: At this point, the beer contains alcohol, but not much carbon dioxide. The brewer has a few options to increase carbon dioxide levels. The most common approach by large-scale brewers is force carbonation, via the direct addition of CO2 gas to the keg or bottle. Smaller-scale or more classically-minded brewers will add extra (*"priming"*) sugar or a small amount of newly fermenting wort (*"kräusen"*) to the final vessel, resulting in a short refermentation known as *"cask-"* or *"bottle conditioning"*.

After brewing, the beer is usually a finished product. At this point the beer is kegged, casked, bottled, or canned. Unfiltered beers may be stored for further fermentation in conditioning tanks, casks or bottles to allow smoothing of harsh alcohol notes, integration of heavy hop flavours, and/or the introduction of oxidised notes such as wine or sherry flavours. Some beer enthusiasts consider a long conditioning period attractive for various strong beers such as Barley wines. There are some beer cafes in Europe, such as Kulminator in Antwerp, which stock beers aged ten years or more. Aged beers such as *Bass Kings Ale* from 1902, *Courage Imperial Russian Stout* and *Thomas Hardys Ale* are particularly valued.

Ingredients

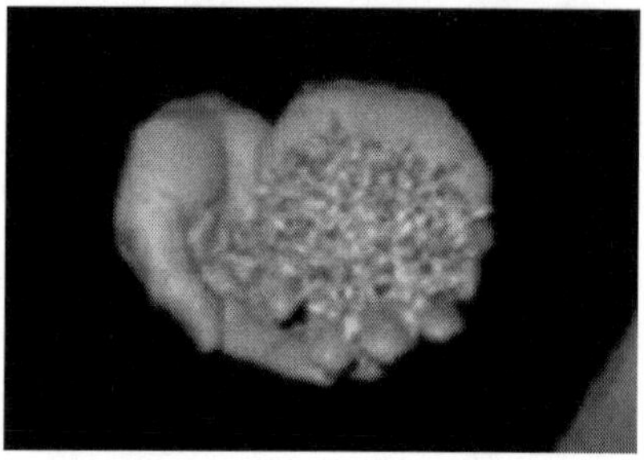

Fig. Malted Barley Before Roasting

The basic ingredients of beer are water, a fermentable starch source, such as malted barley, and yeast. It is common for a flavouring to be added, the most popular being hops. A mixture of starch sources may be used, with the secondary starch source, such as corn, rice and sugar, often being termed an adjunct, especially when used as a lower cost substitute for malted barley.

- *Water*: Beer is composed mainly of water, which when heated is known as brewing liquor. The characteristics of the water have an influence on the character of the beer. Although the effect of, and interactions between, various dissolved minerals in brewing water is complex, as a general rule, hard water is more suited to dark beer such as stout, while very soft water is more suited for brewing pale ale and pale lager.

- *Starch source*: The most common starch source is malted cereal. And among malts, barley malt is the most widely used owing to its high amylase content, a digestive enzyme which facilitates the breakdown of the starch into sugars. However, depending on what can be cultivated locally, other malted and unmalted grains may be used, including wheat, rice, oats, and rye, and less frequently, maize and sorghum. Malt is formed from grain by soaking it in water, allowing it to start to germinate, and then drying the germinated grain in a kiln. Malting the grain produces the enzymes that will eventually convert the starches in the grain into fermentable sugars. Different roasting times and temperatures are used to produce different colours of malt from the same grain. Darker malts will produce darker beers. Two or more types of malt may be combined.

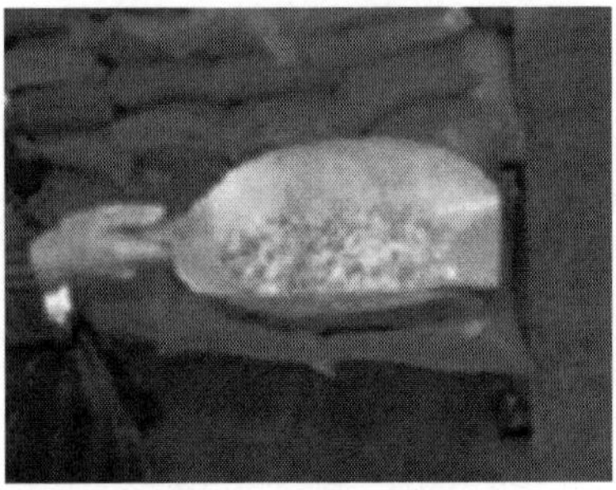

Fig. Crushed Hops

- *Flavourings*: Hops have commonly been used as a bittering agent in beer since the seventeenth century. Hops contain several characteristics very favourable to beer:
  a. Hops contribute a bitterness that balances the sweetness of the malt,
  b. Hops also contribute aromas which range from flowery to citrus to herbal,
  c. Hops have an antibiotic effect that favours the activity of brewer's yeast over less desirable microorganisms and
  d. The use of hops aids in "head retention", the length of time that a foamy head created by the beer's carbonation agent will last.

The bitterness of commercially-brewed beers is measured on the International Bitterness Units scale. While hop plants are grown by farmers all around the world in many different varieties, there is no major commercial use for hops other than in beer.

- *Yeast*: Is a microorganism that is responsible for fermentation. A specific strain of yeast is chosen depending on the type of beer being produced, the two main strains being ale yeast (*Saccharomyces cerevisiae*) and lager yeast (*Saccharomyces uvarum*), with some other variations available, such as *brettanomyces* and *Torulaspora delbrueckii*. Yeast will metabolise the sugars extracted from the grains, and produce alcohol and carbon dioxide as a result. Before yeast's functions were understood, fermentations were conducted naturally using wild or airborne yeasts; although a few styles such as lambics still rely on this ancient method, most modern fermentations are conducted using pure yeast cultures.
- *Clarifying agent*: Some brewers add one or more clarifying agents to beer that are not required to be published as ingredients.

Common examples of these include Isinglass finings, obtained from swimbladders of fish; kappa carrageenan, derived from seaweed; Irish moss, a type of red alga; and gelatin. Since these ingredients may be derived from animals, those concerned with the use or consumption of animal products should obtain specific details of the filtration process from the brewer.

Varieties of Beer

Though there are only a few distinct types of beer, there are many different names and *style* labels that attempt to categorise beers by overall flavour and, occasionally, origin. The British beer writer Michael Jackson wrote about beers from around the world in his 1977 book *The World Guide To Beer* and organised them into local style groups based on local information. This book had an influence on craft and homebrewers in United States who developed an intricate system of categorising beers which is exemplified by the Beer Judge Certification Program. Outside of North America beer is mainly categorised by strength and/or colour.

Categorising by Yeast

A common method of categorising beer is by the behaviour of the yeast used in the fermentation process. In this method of categorising, those beers which use a fast acting yeast which leaves behind residual sugars are termed ales, while those beers which use a slower and longer acting yeast which removes most of the sugars leaving a clean and dry beer are termed lagers.

Ale

Fig. Cask Ales

A modern ale is commonly defined by the strain of yeast used and the fermenting temperature. Ales are normally brewed with top-fermenting yeasts, though a number of British brewers, including Fullers and Weltons, use ale yeast strains that have less pronounced top-fermentation characteristics. The important distinction for ales is that they are fermented at higher temperatures and thus ferment more quickly than "lager"-style beers.

Ale is typically fermented at between (15 and 24°C, 60 and 75°F). At these temperatures, yeast produces significant amounts of esters and other secondary flavor and aroma products, and the result is often a beer with slightly "fruity" compounds resembling but not limited to apple, pear, pineapple, banana, plum, or prune. Typical ales have a sweeter, fuller body than "lagers".

Stylistic differences between some ales and lagers can be difficult to categorise. Steam beer, Kölsch and some modern British Golden Summer Beers are seen as hybrids, using elements of both lager and ale production. Baltic Porter and Bière de Garde may be produced by either lager or ale methods or a combination of both. However, commonly, lager is perceived to be cleaner tasting, dryer and lighter in the mouth than ale.

Lager

Fig. A Glass of Lager

Lager is the English name for bottom-fermenting beers of Central European origin, though the term is not used there. They are the most commonly-consumed beer in the world. The name comes from the German *lagern* ("to store").

Lager yeast is a bottom-fermenting yeast, and typically undergoes primary fermentation at 7-12°C (45-55°F) (the "fermentation phase"), and then is given a long secondary fermentation at 0-4°C (30-40°F) (the "lagering phase"). During the secondary stage, the lager clears and mellows. The cooler conditions also inhibit the natural production of esters and other byproducts, resulting in a "crisper" tasting beer.

Modern methods of producing lager were pioneered by Gabriel Sedlmayr the Younger, who perfected dark brown lagers at the Spaten Brewery in Bavaria, and Anton Dreher, who began brewing a lager, probably of amber-red color, in Vienna in 1840–1841. With modern improved yeast strains, most lager breweries use only short periods of cold storage, typically 1–3 weeks.

The *lagering* phase is not restricted to *lager* beers. In Germany, all beers are stored at low temperatures before consumption; in the British tradition, the practice of Cold Conditioning is similar in nature.

Spontaneous Fermentation

These are beers which use wild yeasts, rather than cultivated ones. Many of these are not related to brewer's yeast (*saccharomyces*), and may have significant differences in aroma and sourness.

Categorising by colour or malt

Another common method of categorising beer is by colour or malts. The colour of a beer is determined by the malt. The most common colour is a pale amber produced from using pale malts. *Pale ale* is a term used for beers made from malt dried with coke. Coke had been first used for roasting malt in 1642, but it wasn't until around 1703 that the term *pale ale* was first used. In terms of volume, most of today's beer is based on the pale lager brewed in 1842 in the town of Plzeò, in the Czech Republic. The modern Pilsner lager is light in colour and high in carbonation, with a strong hop flavour and an alcohol by volume content of around 5%. The Pilsner Urquell and Heineken brands of beer are typical examples of pale lager, as are the American brands Budweiser, Coors, and Miller. Very dark beers , such as stout, use very dark malts.

Draught and keg Beers

Fig. Draught Beer Keg Fonts at the Delirium Café in Brussels

Draught beer from a pressurised keg is the most common dispense method in bars around the world. A metal keg is pressurised with carbon dioxide (CO2) gas which drives the beer to the dispensing tap or faucet. Some beers, notably stouts, such as Guinness and "Smooth" bitters, such as Boddingtons, may be served with a nitrogen/carbon dioxide mixture. Nitrogen has fine bubbles, producing a dense head and a creamy mouthfeel.

In the 1980s Guinness introduced the beer widget, a nitrogen pressurised ball inside a can which creates a foamy head. The words "draft" and "draught" are often used as marketing terms to describe canned or bottled beers containing a beer widget.

Cask Ales

Fig. Schlenkerla Rauchbier Direct from the Cask

Cask ales are unfiltered and unpasteurised. When the landlord feels the beer has settled, and he is ready to serve it, he will knock a soft spile into a bunghole on the side of the cask. The major difference in appearance between a keg and a cask is the bunghole. A keg does not have a bunghole on the side.

The soft spile in the bunghole allows gas to vent off. This can be seen by the bubbles foaming around the spile. The landlord will periodically check the bubbles by wiping the spile clean and then watching to see how fast the bubbles reform. There still has to be some life in the beer otherwise it will taste flat, but too much life and the beer will taste hard or fizzy. When the beer is judged to be ready, the landlord will replace the soft spile with a hard one (which doesn't allow air in or gas out) and let the beer settle for 24 hours.

He will also knock a tap into the end of the cask. This might simply be a tap if the cask is stored behind the bar. The beer will then be served simply under gravity pressure: turn on the tap, and the beer comes out. But if the cask is in the cellar, the beer needs to travel via tubes, or beer lines, up to the bar area using a beer engine.

Bottle Conditioned Beers

Bottle conditioned beers are unfiltered and unpasteurised. It is usually recommended that the beer is poured slowly, leaving any yeast sediment at the bottom of the bottle. However, some drinkers prefer to pour in the yeast, and this practice is customary with wheat beers. Typically when serving a hefeweizen 90% of the contents is poured and the remainder swirled to dissolve the sediment before pouring it into the glass.

Beer Culture

Fig. Gambrinus - King of Beer

Beer in a Social Context

Various social traditions and activities are associated with beer drinking, such as playing cards, darts or other games; or visiting a series of different pubs in one evening. Consumption in isolation and excess may be associated with people "drowning their sorrows," while drinking in excess in company may be associated with binge drinking.

Beer Around the World

Beer is consumed in countries all over the world. There are breweries in Middle Eastern countries such as Iraq and Syria as well as African countries and remote countries such as Mongolia.

Serving

Glassware

An appropriate glass is considered desirable by some beer drinkers. Some drinkers of beer may sometimes drink straight from the bottle or can, while others may pour their beer into a vessel before imbibing. Drinking out of a bottle inhibits aromas picked up by the nose, so if a drinker wishes to appreciate a beer's aroma, the beer is first poured into a glass, mug, tankard, or stein. Some breweries produce glassware intended for their own beers. Some aficionados claim that the shape and material of the vessel influences the perception of the aroma and the way in which the beer settles, similar to claims by drinkers of brandy or cognac.

Some drinkers in Britain prefer their ale to be served in pewter tankards, while in Europe it is common for glasses to be rinsed just before beer is poured into them. While glass is completely nonporous, its surface can retain oil from the skin, aerosolized oil from nearby cooking, and traces of fat from food. When these oils come in contact with beer there is a significant reduction in the amount of head (foam) that is found on the beer, and the bubbles will tend to stick to the side of the glass rather than rising to the surface as normal.

Temperature

The temperature of a beer has an influence on a drinker's experience. Colder temperatures start to inhibit the chemical senses of the tongue and throat, which narrow down the flavour profile of a beer, allowing fully attenuated beers such as Pilsners and Pale lagers to be enjoyed for their crispness, but preventing the more rounded flavours of an ale or a stout from being perceived. While there are no firmly agreed principles for all cases, a general approach is that lighter coloured beers, such as Pale lagers, are usually enjoyed cold (40-45F/4-7C), while dark, strong beers such as Imperial Stouts are often enjoyed at cellar temperature (54-60F/12-16C) and then allowed to warm up in the room to individual taste. Other beers should be served at temperatures between these extremes.

Pouring

The pouring process has an influence on a beer's presentation. The rate of flow from the tap or other serving vessel, tilt of the glass, and position of the pour (in the center or down the side) into the glass all influence the end result, such as the size and longevity of the head, lacing (the pattern left by the head as it moves down the glass as the beer is drunk), and turbulence of the beer and its release of carbonation.

Unfiltered bottled beers may be served with the addition of the remaining yeast at the bottom of the bottle to add both flavour and colour.

Rating Beer

Rating beer is a recent craze that combines the enjoyment of beer drinking with the hobby of collecting. People drink beer and then record their scores and comments on various internet websites. This is a worldwide activity and people in the USA will swap bottles of beer with people living in New Zealand and Russia. People's scores may be tallied together to create lists of the most popular beers in each country as well as the most highly rated beers in the world.

Health Effects

Beer contains alcohol which has a number of health risks and benefits. However, beer includes a wide variety of other agents that are currently undergoing scientific evaluation. Nutritionally, beer can contain significant amounts of magnesium, selenium, potassium, phosphorus, biotin, and B vitamins. Typically, the darker the brew, the more nutrient dense.

A 2005 Japanese study found that non-alcoholic beer may possess strong anti-cancer properties. Another study found nonalcoholic beer to mirror the cardiovascular benefits associated with moderate consumption of alcoholic beverages. It is considered that over-eating and lack of muscle tone is the main cause of a beer belly, rather than beer consumption.

The Strongest Beers In The World

Beer strength varies by local custom. British ale tends to average 4.4% alcohol by volume (abv). Belgian beers tend to average 8% abv. The strength of the typical global pale lager is 5% abv. Typical brewers yeast cannot reproduce (and thereby cannot produce alcohol) above 12% abv; however, in the 1980s the Swiss brewery Hurlimann developed a yeast strain which could get as high as 14% for their Samichlaus beer. Since then breweries have experimented with using champagne yeasts, continually pushing up the strength.

Samuel Adams reached 20% abv with Millennium and then surpassed that amount to 25.6% abv with Utopias. The strongest beer sold in Britain was Dogfish Head's World Wide Stout, a 21% abv stout which was available from UK Safeways in 2003. In Japan in 2005, the Hakusekikan Beer Restaurant sold an eisbock, strengthened through freezing, believed to be 28% abv. The beer that is considered to be the strongest yet made is Hair of the Dog's Dave - a 29% abv barley wine made in 1994.

Gluten free Beer

Around one in a hundred people (known as coeliacs) cannot drink beer due to a severe reaction to gluten, specifically the gliadin found in wheat beer but more frequently, the hordein that is the "gluten" of barley. The international resource for gluten free beer is Glutenfreebeerfestival.com and this site reviews all "gluten free" beers, and arranges festivals where coeliacs may drink beer like anyone else.

PRODUCTS OF MIXED FERMENTATIONS

Most traditional fermented food products are made by a complex interaction of different micro-organisms.

Vinegars

Vinegar is the product of a mixed fermentation of yeast followed by acetic acid bacteria. Vinegar, literally translated as sour wine, is one of the oldest products of fermentation used by man. It is the acetic acid produced by the fermentation of alcohol (ethanol) which gives the characteristic flavour and aroma to vinegar. It can be made from almost any fermentable carbohydrate source, for example fruits, vegetables, syrups and wine. The basic requirement for vinegar production is a raw material that will undergo an alcoholic fermentation. Apples, pears, grapes, honey, syrups, cereals, hydrolysed starches, beer and wine are all ideal substrates for the production of vinegar. To produce a high quality product it is essential that the raw material is mature, clean and in good condition.

Indigenous vinegars can be made quite simply by the spontaneous fermentation of a fruit or alcohol. All that is necessary is an alcoholic substrate, strains of acetic-acid forming bacteria (*acetobacter*) and oxygen to enable the oxidation of alcohol. However, this process is very slow and vinegars produced by this method tend to be of inferior quality. Controlled fermentation conditions produce a more acceptable product. A wide range of raw materials can be made into vinegar.

Coconut Water Vinegar

Location of Production

Throughout Asia particularly the Philippines and Sri Lanka

Product Description

A clear liquid with a distinctive acetic acid taste with a hint of a coconut flavour.

Raw Material Preparation

Coconut water is a waste product, which is produced in appreciable quantities in the Philippines, Sri Lanka Thailand and other countries. Its conversion into vinegar therefore presents an attractive option for decreasing wastage and producing a valuable product.

Processing

Coconut water is a good base for vinegar, but its sugar content is too low (only about 1%). Sugar needs to be added to bring the level of sugar up to 15%. After the addition of sugar, the coconut juice is allowed to ferment for about seven days, during which time the sugar is converted to alcohol. An alternative method is to pasteurise the coconut water and sugar mixture and add yeast.

After this initial fermentation, strong vinegar (10% v/v) is added to stimulate the growth of acetic acid bacteria and discourage further yeast fermentation. The acetic acid fermentation takes approximately one month, yielding a vinegar with approximately 6% acetic acid. The fermentation will take less time than this if a generator is used. After fermentation, the vinegar must be stored in anaerobic conditions to prevent spoilage by the oxidation of acetic acid. (Steinkraus, 1996) Clarification can be achieved by stirring with a well beaten egg white, heating until the egg white coagulates and filtering (Anon).

Pineapple Peel Vinegar

Location of Production

Latin America and Asia

Product Description

This product enables the utilisation of pineapple peels, which are usually discarded during the processing or consumption of the fruit. The product has a distinct, very light pineapple flavour and has the same uses as any commercial vinegar.

Raw Material Preparation

The peels should be from very well washed ripe pineapples (damaged, rotten or infected fruits should not be used as a source of peels). Use only the peels, not the leaves or stems. The water used should be potable water, boiled if necessary. All the equipment should be well cleaned, as well as the bottles, which should also be steam-sterilised before use.

Processing

The peels should be cut into thin strips and put into clay or pewter pots. Aluminium or iron pots should not be used. Sugar and clean water are added. Each pot is then inoculated and covered with a clean cotton cloth, held around the pot with an adhesive tape, to prevent contamination by insects or dust.

The inoculated pineapple is fermented at room temperature (about 20-22°C) for about eight days. The acidity should be checked daily. The water level should be maintained during this period. The product should be increasingly acid and by the eighth day it should have the required concentration of 4 per cent acetic acid in vinegar. If higher acidity is desired the product is left to ferment for another one or two days. The development of acidity should be checked by tasting the product during fermentation. The residual bacteria removed may be reused as a residue inoculum two or three times more.

The traditional process may be improved by a two-stage fermentation in which alcohol is first formed by yeast (*Saccharomyces cerevisiae*) and the 'must' is then inoculated with acetic acid bacteria (*Acetobacter pasteurianus*). In outline, the process involves liquidising the peels and diluting with water (water:pulp is 4:1), adjusting the pH to 4.0 using sodium bicarbonate and adding yeast nutrient (ammonium phosphate) at 0.14g per litre.

A starter culture is added at 2.7g per litre and the fermentation allowed to take place at 25°C for two days. The 'must' is then filtered and inoculated with acetic acid bacteria and allowed to ferment for eleven days with aeration of the 'must'. Other parts of the process are similar. Additional equipment includes a pH meter, refractometer, liquidiser, fermentation locks and equipment for preparing the starter cultures (Fellows, 1997).

Packaging and Storage

The vinegar is bottled in clean glass bottles and stored in a cool dark place.

Palm Wine Vinegar

Palm wine vinegar is a produced across West Africa. It is a vinegar containing about 4% acetic acid, produced from the oxidation of palm wine. It is mainly consumed by people in urban areas as a salad dressing and meat tenderiser, although it also has medicinal uses and is valued in certain rituals. Palm wine is fermented using the same process as for grape wine vinegar – the oxidation of alcohol to acetic acid. The spontaneous process takes about four days. The optimum fermentation temperature is 30º C.

Coconut Toddy Vinegar

Coconut toddy vinegar is produced throughout South Asia particularly Sri Lanka. It is a clear liquid with a strong acetic acid flavour and a hint of coconut flavour. The fresh toddy is strained, prior to allowing yeast fermentation to occur naturally for 48 to 72 hours. The yeast cells and debris are then removed by progressive sedimentation. After two to four weeks of settling the fermented toddy is placed in barrels. The alcohol is then converted into acetic acid by acetic acid bacteria which are naturally present. The process can be hastened by adding vinegar as a starter. The fermented toddy is converted into vinegar in about three months. Ageing for six months, results in a pleasantly flavoured final product.

Nipa Palm Vinegar

In East Asia particularly Papua New Guinea a vinegar is made from the sap of the *Nipa* palm (*Nypa fruticans*)

Quick process Pickles

Quick process pickles are easy to make but do not really constitute a fermented food product. For this technique, vegetables are soaked in a low salt solution for a few hours. They are then drained and placed in a container. The container is filled with a hot vinegar and spice mixture or a hot oil and spice mixture (Kordylas, 1990). There are hundreds of different recipes utilising locally available fruit and vegetables. For instance the book "Pickles of Bangladesh" has recipes for mango sour pickle, sliced mango pickle, sweet olive pickle, hot olive pickle, sweet tamarind pickle, chalta pickle and green chilli pickle .

Cocoa Products

Location of Production

Africa, Asia and Latin America particularly Cote d'Ivoire, Ghana, Indonesia and Brazil.

Product Description

A fine brown powder with the characteristic taste of cocoa. It is a major ingredient in the confectionery and bakery industries. The product has a short shelf life. "Drinking chocolate" is a mixture of cocoa powder and sugar.

Raw Material Preparation

Cocoa beans are the seeds of the cocoa plant (*Theobroma cacao*). Cocoa pods are cut from the cocoa tree. The pods are cut and the beans removed. Only fully ripe and undamaged beans should be selected. It is important that the beans are processed quickly.

Processing

It was formerly believed that cocoa beans were fermented to remove the adhering pulp (Wood, 1990). However a good flavour in the final cocoa or chocolate is dependent on good fermentation. Fermentation is carried out in a variety of ways but all depend on heaping a quantity of fresh beans with their pulp and allowing micro-organisms to produce heat (Beckett, 1988).

The majority of beans are fermented in heaps although better results are obtained using boxes, which result in a more even fermentation.

Fermentation lasts from five to six days. During the first day the adhering pulp is liquified and drains away with the temperature rising steadily. The initial alcoholic fermentation gives way to acetification. This and other chemical changes cause the temperature to rise in excess of 50oC. The beans die. It was thought in the past that death was mainly due to increasing temperature.

It is now known that acetic acid at a concentration of 1 percent in the bean is the cause of death and that it is only enhanced by heat, lactic acid and ethanol. The pH value of the cotyledon drops from 6.45 to 4.5 over 120 hours and that during the same period the acetic acid content increased from 0 to 1.36 percent, while the lactic acid content increased from 0.005 to 0.12 percent. When the bean dies maceration of the tissue takes place, allowing enzymes and substrate to mix freely.

The possible substrates for enzymes are carbohydrates, lipids, phenolics and amino acids. In addition it is known that the bacteria can metabolise alcohols and organic acids of various kinds. The changing chemical picture is complex. Possible major substrates for micro-organisms are carbohydrates, lipids, phenolics and amino acids. Unlike some flavours and aromas, that of chocolate is not attributable to a single compound (Carr, 1985) (Minifie, 1980).

During fermentation the external appearance of the beans changes. At first they are pinkish with a covering of white mucilage. Gradually the colour darkens and the mucilage disappears. The beans on the surface are always darker than those deeper in the heap or box, indicating that the colour change is oxidative. As the beans are mixed, their colour becomes a more uniform orange-brown and they are only slightly sticky. At this stage they are ready for drying.

The beans need to be dried to a moisture content of less than 7.5%. The beans are dried by either being spread out in the sun in layers a few centimetres thick or in artificial dryers. There are numerous types of dryers but it is important that any smoky products of combustion do not come in contact with the beans otherwise taints will appear in the final product. The beans are cleaned to remove the extraneous matter. Cocoa beans consist of an outer skin that needs to be removed and inner "nib". The shell is sometimes removed before roasting and sometimes after roasting.

For cocoa powder roasting temperatures of 120 to 150° C are used. There are many designs of roasters: both batch and continuous systems. The operation is controlled so that the cocoa is heated to the required temperature without burning the shell or the cotyledon. The heat is applied evenly over a long period of up to 90 minutes to produce even roasting. The bean must not be contaminated with any combustion products from the fuel used and provision must be made for the escape of any volatile acids, water vapour and decomposition products of the bean (Wood, 1980) (Cakebread, 1975). After roasting the beans are cooled quickly to prevent scorching.

The roasted nibs are ground into a powder in a plate mill. The resulting powder is sieved through fine silk, nylon or wire mesh. To produce cocoa powder, some of the cocoa butter needs to be removed. With low fat cocoa powder, more than 90% of the cocoa butter is removed. With medium fat cocoa powder, more than 78% of the cocoa butter has been removed. Finally high fat cocoa powder has less than 78% of the cocoa butter removed. Extrusion, expeller, or screw presses are used in the cocoa industry to remove the cocoa The cake from the mill is ground in a hammer mill to produce the cocoa powder.

Packaging and Storage

Cocoa powder is hygroscopic (picks up moisture from the air) and should be protected, especially in humid climates. Lidded tins or sealed polythene bags should be used.

Chocolate

Location of Production

Throughout Africa, Asia and Latin America.

Product Description

A brown solid oily product with the characteristic taste of chocolate. It is a major ingredient in the confectionery and bakery industries.

Preparation of Raw Materials

Cocoa beans are the seeds of the cocoa plant (*Theobroma cacao*). Cocoa pods are cut from the trees, and the beans are removed from the pods. Only fully ripe and undamaged beans should be selected. It is important that the beans are processed quickly.

Processing

Fermentation, drying and cleaning of the beans have been described in Section 7.2.1. For cocoa butter production the roasting temperatures are 100° C to 104° C There are many designs of roasters: both batch and continuous systems.

The operation is controlled so that: the cocoa is heated to the required temperature without burning the shell or the cotyledon The heat is applied evenly over a long period of up to 90 minutes to produce even roasting; the nib must not be contaminated with any combustion products from the fuel used and provision must be made for the escape of any volatile acids, water vapour and decomposition products of the nib (Wood, 1980) After roasting the beans are cooled quickly to prevent scorching Roasting will have already loosened the shell. The beans are then lightly crushed with the object of preserving large pieces of shell and nib and avoiding the creation of small particles and dust. The cocoa bean without its shell is known as a "cocoa nib". The valuable part of the cocoa bean is the nib, the outer shell being a waste material of little value.

Alkalization is a treatment that is sometimes used before and sometimes after grinding to modify the colour and flavour of the product. This was developed in the Netherlands in the last century and is sometimes known as "Dutching". This involves soaking the nib or the cocoa mass in potassium or sodium carbonate. By varying the ratio of alkali to nib, a wide range of colours of cocoa powder can be produced (Glossop, 1993). Complete nib penetration may take an hour. After alkalization the cocoa needs to be dried slowly.

The cocoa nib is ground into "cocoa liquor" (also known as "unsweetened chocolate" or "cocoa mass"). The grinding process generates heat and the dry granular consistency of the nib is turned into a liquid as the high amount of fat contained in the nib melts (Gates, 1990). There are various pre-treatments to develop the flavour of the cocoa mass with and without reaction solutions. These include the "Luwa thin-layer evaporator", "Petzomat thin-layer process", "Cocovap process", "Lehman KFA process" and "Carle-Montanari process" (Beckett, 1988) .

Extrusion, expeller, or screw presses are used in the cocoa industry for the production of cocoa butter from whole beans, and mixtures of fine nib dusts, small nibs, and immature beans. Research in the Kerala Agricultural University has led to develop a suitable pressure device capable of separating cocoa butter from ground cocoa mass ideally suitable for small scale manufacturers (Ganeshan, 1990). In Peru a simple screw press is used to extract cocoa butter from beans. The crude cocoa butter is filtered through cloth and allowed to solidify.

To produce plain chocolate, cocoa mass is mixed with sugar and sufficient cocoa butter to enable the chocolate to be moulded. The ratio of mass to sugar varies according to the national taste. The mixture is ground to such a degree that the chocolate is smooth to the palate. At one time this was done by a lengthy process in melengeurs - heavy granite rollers in a revolving granite bed - but nowadays grinding is done in a series of rolls.

The chocolate is then "Conched". This may last for several hours. The chocolate is heated, this helps to drive off volatile acids, thereby reducing acidity when present in the raw bean, and the process finishes the development of flavour and makes the chocolate homogeneous (Wood, 1980)). Similar processes are involved in the manufacture of milk chocolate. The milk is added in various ways either in powder form to the mixture of mass, sugar and cocoa butter, or by condensing first with sugar, adding the mass and drying this mixture under vacuum. The product is called 'crumb' and this is ground and conched in a similar manner to plain chocolate. After conching the chocolate has to be tempered before it is used for moulding or for enrobing confectionery centres. Tempering involves cooling and reaching the right physical state for rapid setting after moulding or enrobing.

Coffee

Location of Production

Throughout Africa, Asia and Latin America particularly Brazil, Colombia, Indonesia, Mexico and Cote d'Ivoire.

Product Description

A fine dark brown powder made from roasted coffee beans. Brewed with boiling water and consumed as a drink.

Raw Material Preparation

Coffee beans are harvested from two plants *Coffea arabica and Coffea canephora* variety *robusta*. Only ripe berries should be used in coffee production. Berries can be placed in water so that immature berries which float can be identified and discarded.

Processing

Dry processing is the simpler of the two processing methods and is popular in Brazil for the processing of *Robusta* coffee and in Sri Lanka for processing *Arabica* coffee. The coffee cherries are dried immediately after harvest by sun drying on a clean dry floor or on mats. The dried berry is then hulled to remove the pericarp. This can be done by hand using a pestle and mortar or in a mechanical huller. The mechanical hullers usually consist of a steel screw, the pitch of which increases as it approaches the outlet so removing the pericarp. The hulled coffee is cleaned by winnowing.

Wet processing involves squeezing the berry in a pulping machine or pounding in a pestle and mortar to remove the outer fleshy material (mesocarp and exocarp) and leave the bean covered in mucilage. This mucilage is removed by fermentation. Fermentation involves placing the beans in plastic buckets or tanks and allowing them to sit, until the mucilage is broken down. Natural enzymes in the mucilage and yeasts and bacteria in the environment work together to break down the mucilage. The coffee should be stirred occasionally and every so often a handful of beans should be tested by washing in water. If the mucilage can be washed off and the beans feel gritty rather than slippery, the beans are ready.

There is much debate about the fermentation of coffee beans. Some researchers feel that the mucilage breakdown is caused by enzymatic breakdown. For instance Wellman has stated that enymatic fermentation starts immediately the beans have been squeezed from the fresh berries. If these 'pulped' beans are piled up or put in a container and protected from any bacterial or other contamination the fermentation will progress.

After a number of hours the enzymes of the pulp will have acted on the torn tissues, gorged with starches, sugars and pectins, in such a manner that, without any microbial intervention, the remaining pulp will be easily detached from the beans and washed off in water (Wellman, 1961). However most investigators acknowledge the necessity for the presence of micro-organisms for the depectinisation of the beans.

The following micro-organisms have been isolated: *Leuconostoc mesenteroides; Lactobacillus plantarum; Lactobacillus brevis, Streptococcus faecalis, Aerobacter (Enterobacter)* and *Escherichia,* pectinolytic species of *Bacillus, Saccharomyces marscianus, S. bayanus and* a *Flavobacterium* sp., *Erwinia dissolvens, Fusarium spp, Aspergillus spp* and *Penicillium. (Pedersen and Breed)* (Vaughn *et al.,* 1958) (Hilmer *et al.,*1965), (Agate and Bhat, 1966).

The beans should then be washed immediately as 'off' flavours develop quickly. To prevent cracking the coffee beans should be dried slowly to 10% moisture content (wet basis). Drying should take place immediately after to prevent 'off' flavours developing. The same drying methods can be used for this as for the dry processed coffee. After drying the coffee should be rested for 8 hours in a well ventilated place. The thin parchment around the coffee is removed either by hand, in a pestle and mortar or in a small huller. The hulled coffee is cleaned by winnowing.

The final flavour of the coffee is heavily dependent on how the beans are roasted. Roasting is a time temperature dependent process. The roasting temperature needs to be about 200° C. The degree of roast is usually assessed visually.

One method is to watch the thin white line between the two sides of the bean, when this starts to go brown the coffee is ready. As preferences vary considerably from region to region, a lot of research will need to be done to find the locally acceptable degree of roast. Coffee beans can be roasted in a saucepan as long as they are continually stirred. A small improvement is made by roasting the coffee in sand, as this provides a more even heat. A roaster will produce a higher quality product.

Grinding is a means of adding value to a product. However, it is fraught with difficulties. It is easy to make an assessment of an intact bean, while a ground product presents some difficulty. The fear of adulteration and the use of low quality produce is justified. Because of this there is a great deal of market resistance to ground coffee. This market resistance can only be overcome by consistently producing a good product. There are basically two types of grinders - manual grinders and motorised grinders (anon, 1995).

Packaging and Storage

Roasted beans can be stored in sacks. Milled beans need to be packaged quickly to prevent the loss of volatile flavour components. The packaging material should be airtight. Polythene is not suitable as it is a low barrier to loss of aroma (Fellows, 1997).

Other Mixed Fermentation Products

Vanilla

Vanilla is produced in Madagascar, Indonesia and various South Pacific islands. It is a dark brown pod about 20 cm in length. Vanilla is produced by fermenting the pods of the orchids of the genus vanilla. The pods are first sun dried for 24 to 36 hours and then blanched in hot water (65° C) for two to three minutes. The pods are then fermented in boxes and dried again.

Tabasco

Tabasco sauce is made in Mexico and Guatemala. The chilli pods are harvested, ground into a paste and placed in a container with salt. The hot and fiery sauce develops.

Tea

In the production of tea, there is a process referred to as fermentation. However microbial activity is not involved in the so-called 'fermentation' of tea. The chemical changes are effected by enzymes alone. Fermentation rooms are used where moisture and temperature can be controlled. During fermentation even further darkening of the leaf occurs and the typical aroma develops. By subjective judgement of the aroma's intensity the period necessary for completion is gauged .

7

Dairy Products, Spoilage Microbiology and Types of Spoilage Microorganisms

TYPES OF SPOILAGE MICROORGANISMS

PSYCHROTROPHS

Psychrotrophic microorganisms represent a substantial percentage of the bacteria in raw milk, with pseudomonads and related aerobic, Gram-negative, rod-shaped bacteria being the predominant groups. Typically, 65–70 per cent of the psychrotrophs isolated from raw milk are *Pseudomonas* species. Important characteristics of pseudomonads are their abilities to grow at low temperatures (3–7oC) and to hydrolyze and use large molecules of proteins and lipids for growth. Other important psychrotrophs associated with raw milk include members of the genera *Bacillus, Micrococcus, Aerococcus,* and *Lactococcus* and of the family Enterobacteriaceae. Pseudomonads can reduce the diacetyl content of buttermilk and sour cream, thereby leading to a "green" or yogurt- like flavour from an imbalance of the diacetyl to acetaldehyde ratio.

For cottage cheese, the typical pH is marginally favourable for the growth of Gram-negative psychrotrophic bacteria, with the pH of cottage cheese curd ranging from 4.5 to 4.7 and the pH of creamed curd being within the more favourable pH range of 5.0–5.3. The usual salt content of cottage cheese is insufficient to limit the growth of contaminating bacteria; therefore, psychrotrophs are the bacteria that normally limit the shelf life of cottage cheese. When in raw milk at cell numbers of greater than 106 CFU/ml, psychrotrophs can decrease the yield and quality of cheese curd.

COLIFORMS

Like psychrotrophs, coliforms can also reduce the diacetyl content of buttermilk and sour cream, subsequently producing a yogurt-like flavour. In cheese production, slow lactic acid production by starter cultures favours the growth and production of gas by coliform bacteria, with coliforms having short generation times under such conditions. In soft, mold-ripened cheeses, the pH increases during ripening, which increases the growth potential of coliform bacteria.

LACTIC ACID BACTERIA

Excessive viscosity can occur in buttermilk and sour cream from the growth of encapsulated, slime-producing lactococci. In addition, diacetyl can be reduced by diacetyl reductase produced in these products by lactococci growing at 7°C, resulting in a yogurt-like flavour. Heterofermentative lactic acid bacteria such as lactobacilli and *Leuconostoc* can develop off-flavours and gas in ripened cheeses. These microbes metabolize lactose, subsequently producing lactate, acetate, ethanol, and CO2 in approximately equimolar concentrations.

Their growth is favoured over that of homofermentative starter culture bacteria when ripening occurs at 15°C rather than 8°C. When the homofermentative lactic acid bacteria fail to metabolize all of the fermentable sugar in a cheese, the heterofer-mentative bacteria that are often present complete the fermentation, producing gas and off-flavours, provided their populations are 106 CFU/g. Residual galactose in cheese is an example of a substrate that many heterofermentative bacteria can metabolize and produce gas. Additionally, facultative lactobacilli can cometabolize citric and lactic acids and produce CO2.

Catabolism of amino acids in cheese by non-starter culture, naturally occurring lactobacilli, propionibacteria, and *Lactococcus lactis* subsp. *lactis* can produce small amounts of gas in cheeses. Cracks in cheeses can occur when excess gas is produced by certain strains of *Streptococcus thermophilus* and *Lactobacillus helveticus* that form CO2 and 4-aminobutyric acid by decarboxylation of glutamic acid. Metabolism of tyrosine by certain lactobacilli causes a pink to brown discolouration in ripened cheeses. This reaction is dependent on the presence of oxygen at the cheese surface. The racemic mixture of L(+) and D(–)-lactic acids that forms a white crystalline material on surfaces of Cheddar and Colby cheeses is produced by the combined growth of starter culture lactococci and non-starter culture lactic acid producers. The latter racemize the L(+) form of the acid to the L(–) form, which form crystals.

FUNGI

Yeasts can grow well at the low pH of cultured products such as in buttermilk and sour cream and can produce off-flavours described as fermented or yeasty. Additionally, yeasts can metabolize diacetyl in these products, thereby leading to a yogurt-like flavour. Contamination of cottage cheese with the common yeast *Geotrichum candidum* often results in a decrease of diacetyl content. *Geotrichum candidum* reduced by 52–56 per cent diacetyl concentrations in lowfat cottage cheese after 15–19 days of storage at 4–7°C.

Yeasts are a major cause of spoilage of yogurt and fermented milks in which the low pH provides a selective environment for their growth. Yogurts produced under conditions of good manufacturing practices should contain no more than 10 yeast cells and should have a shelf life of 3–4 weeks at 5°C. However, yogurts having initial counts of >100 CFU/g tend to spoil quickly. Yeasty and fermented off-flavours and gassy appearance are often detected when yeasts grow to 105–106 CFU/g.

Giudici, Masini, and Caggia studied the role of galactose in the spoilage of yogurt by yeasts and concluded that galactose, which results from lactose hydrolysis by the lactic starter cultures, was fermented by galactose-positive strains of yeasts such as *Saccharomyces cerevisiae* and *Hansenula anomala*. The low pH and the nutritional profile of most cheeses are favourable for the growth of spoilage yeasts. Surface moisture, often containing lactic acid, peptides, and amino acids, favours rapid growth.

Many yeasts produce alcohol and CO2, resulting in cheese that tastes yeasty. Packages of cheese packed under vacuum or in modified atmospheres can bulge as a result of the large amount of CO2 produced by yeast. Lipolysis produces short-chain fatty acids that combine with ethanol to form fruity esters. Some proteolytic yeast strains produce sulfides, resulting in an egg odour. Common contaminating yeasts of cheeses include *Candida* spp., *Kluyveromyces marxianus, Geotrichum candidum, Debaryomyces hansenii,* and *Pichia* spp.

Molds can grow well on the surfaces of cheeses when oxygen is present, with the low pH being selective for them. In packaged cheeses, mold growth is limited by oxygen availability, but some molds can grow under low oxygen tension. Molds commonly found growing in vacuum-packaged cheeses include *Penicillium* spp. and *Cladosporium* spp.. *Penicillium* is the mold genus most frequently occurring on cheeses. A serious problem with mold spoilage of sorbatecontaining cheeses is the degradation of sorbic acid and potassium sorbate to *trans*-1, 3-pentadiene, causing an off-odour and flavour described as "kerosene."

Several fungal species, including *Penicillium roqueforti,* are capable of metabolizing this compound from sorbates. Marth, Capp, Hasenzahl, Jackson, and Hussong, who was the first group to study this problem, determined that cheese-spoilage isolates of *Penicillium* spp. were resistant to up to 7,100 ppm of potassium sorbate. Later, Sensidoni, Rondinini, Peressini, Maifreni, and Bortolomeazzi isolated from Crescenza and Provolone cheeses sorbate-resistant strains of *Paecilomyces variotii* and *D. hansenii* (a yeast) that produced *trans*-1, 3-pentadiene, causing offflavours in those products. Cream cheeses are susceptible to spoilage by heat-resistant molds such as *Byssochlamys nivea. Byssochlamys nivea* is capable of growing in reduced oxygen atmospheres, including in atmospheres containing 20, 40, and 60 per cent carbon dioxide with less than 0. 5 per cent oxygen. Once this mold is present in the milk supply, it can be difficult to eliminate during normal processing of cream cheese. Engel and Teuber studied the heat resistance of various strains of *B. nivea* ascospores in milk and cream and determined a *D*-value of 1. 3–2. 4 s at 92°C, depending on the strain. They calculated that in a worstcase scenario of 50 ascospores of the most heat-resistant strain per litre of milk, a process of 24 s at 92°C would result in a 1 per cent spoilage rate in packages of cream cheese.

SPORE-FORMING BACTERIA

Raw milk is the usual source of spore-forming bacteria in finished dairy products. Their numbers before pasteurization seldom exceed 5,000/ml; however, they can also contaminate milk after processing. The most common spore-forming bacteria found in dairy products are *Bacillus licheniformis, B. cereus, B. subtilis, B. mycoides,* and *B. megaterium.* In one study, psychrotrophic *B. cereus* was isolated in more than 80 per cent of raw milks sampled. The heat of pasteurization activates (heat shock) many of the surviving spores so that they are primed to germinate at a favourable growth temperature.

Coagulation of the casein of milk by chymosin-like proteases produced by many of these bacilli occurs at a relatively high pH. Cromie reported that lactose-fermenting *B. circulans* was the dominant spoilage microbe in aseptically packaged pasteurized milk. *Bacillus stearothermophilus* can survive ultra-high-temperature treatment of milk. This bacterium produces acid but no gas, hence causing the "flat sour" defect in canned milk products. If extensive proteolysis occurs during aging of ripened cheeses, the release of amino acids and concomitant increase in pH favours the growth of clostridia, especially *Clostridium tyrobutyricum,* and the production of gas and butyric acid.

FACTORS AFFECTING SPOILAGE

SPOILAGE OF FLUID MILK PRODUCTS

The shelf life of pasteurized milk can be affected by large numbers of somatic cells in raw milk. Increased somatic cell numbers are positively correlated with concentrations of plasmin, a heat-stable protease, and of lipoprotein lipase in freshly produced milk. Activities of these enzymes can supplement those of bacterial hydrolases, hence shortening the time to spoilage.

The major determinants of quantities of these enzymes in the milk supply are the initial cell numbers of psychrotrophic bacteria, their generation times, their abilities to produce specific enzymes, and the time and temperature at which the milk is stored before processing. Several conditions must exist for lipolyzed flavour to develop from residual lipases in processed dairy foods, that is, large numbers (>106 CFU/ml) of lipase producers, stability of the enzyme to the thermal process, long- term storage and favourable conditions of temperature, pH, and water activity.

SPOILAGE OF CHEESES

Factors that determine the rates of spoilage of cheeses are water activity, pH, salt to moisture ratio, temperature, characteristics of the lactic starter culture, types and viability of contaminating microorganisms, and characteristics and quantities of residual enzymes. With so many variables to affect deteriorative reactions, it is no surprise that cheeses vary widely in spoilage characteristics. Soft or unripened cheeses, which generally have the highest pH values, along with the lowest salt to moisture ratios, spoil most quickly.

In contrast, aged, ripened cheeses retain their desirable eating qualities for long periods because of their comparatively low pH, low water activity, and low redox potential. For fresh, raw milk pasta filata cheeses, Melilli determined that low initial salt and higher brining temperature (18 per centC) allowed for greater growth of coliforms, which caused gas formation in the cheese. Factors affecting the growth of the spoilage microorganisms, *Enterobacter agglomerans* and *Pseudomonas* spp. in cottage cheese, were higher pH and storage temperature of the cheese. Some of the spoilage microorganisms were able to grow at relatively low pH values when incubated at 7°C and were able to grow at pH 3. 6 when grown in media at 20°C. Rate of salt penetration into brined cheeses, types of starter cultures used, initial load of spores in the milk used for production, pH of the cheese, and ripening temperature affect the rate of butyric acid fermentation and gas production by *C. tyrobutyricum*.

Fungal growth in packaged cheeses was found to be most significantly affected by the concentration of CO2 in the package and the water activity of the cheese. Cheddar cheese exhibiting yeast spoilage had a high moisture level (39.1 per cent) and a low salt in the moisture-phase value (3.95 per cent). Roostita and Fleet determined that the properties of yeasts that affected the spoilage rate of Camembert and blue-veined cheeses were the abilities to ferment/assimilate lactose, produce extracellular lipolytic and proteolytic enzymes, utilize lactic and citric acid, and grow at 10°C.

SOURCES OF SPOILAGE MICROORGANISMS

CONTAMINATION OF RAW MILK

The highly nutritious nature of dairy products makes them especially good media for the growth of microorganisms. Milk contains abundant water and nutrients and has a nearly neutral pH. The major sugar, lactose, is not utilized by many types of bacteria, and the proteins and lipids must be broken down by enzymes to allow sustained microbial growth. In order to understand the source of many of the spoilage microflora of dairy products, it is best to discuss how milk can first become contaminated, via the conditions of production and processing.

The mammary glands of many very young cows yield no bacteria in aseptically collected milk samples, but as numbers of milkings increase, so do the chances of isolating bacteria in milk drawn aseptically from the teats. The stresses placed on the cow's teats and mammary glands by the very large amounts of milk produced and the actions of the milking machine cause teat canals to become more open and teat ends to become misshapen as time passes. These stresses may open the teat canal for the entry of bacteria capable of infecting the glands.

Environmental contaminants represent a significant percentage of spoilage microflora. They are ubiquitous in the environment from which they contaminate the cow, equipment, water, and milkers' hands. Since milking machines exert about 38 cm (15 in.) of vacuum on the teats during milking, and since air often leaks into the system, bacteria on the surfaces of the cow or in water retained from premilking preparation can be drawn into the milk. Also, when inflation clusters drop to the floor, they pick up microorganisms that can be drawn into the milk. The pumping or agitation of milk supplies the oxygen needed by aerobes for growth and breaks chains and clumps of bacteria. Single cells, having less competition than those in colonies, have the opportunity for more rapid multiplication.

Bacteria recontaminating pasteurized milk originate primarily from water and air in the filling equipment or immediate surroundings and can be resident for prolonged periods of time. In a study performed in Norway and Sweden, Ternstrom, Lindberg, and Molin investigated nine dairy plants and found that five taxa of psychrotrophic *Pseudomonas* spp. were involved in the spoilage of raw and pasteurized milk and that the same strains were recovered from both the raw and pasteurized milk, suggesting that recontamination originated from the raw milk.

Additionally, the investigators found that *Bacillus* spp. (mainly *B. cereus* and *B. polymyxa*) were responsible for spoilage in 77 per cent of the samples that had been spoiled by Gram-positive bacteria. The spoilage *Bacillus* spp. grew fermentatively, and most were able to denitrify the milk, which has implications for cheeses that contain added nitrate/nitrites for protection against clostridia. Sporeforming bacteria are abundant in dust, dairy feed concentrates, and forages; therefore, they are often present on the skin and hair of cattle from which they can enter milk. The presence of sporeformers such as *C. butyricum* in milk has been traced to contaminated silage.

CONTAMINATION OF DAIRY PRODUCTS

Washed curd types of cheeses are especially susceptible to growth of coliforms, so great care must be taken to monitor the quality of water used in these processes. A high incidence of contamination of brine-salted cheeses by yeasts results from their presence in the brines. Many mold species are particularly well adapted to the cheese-making environment and can be difficult to eradicate from a production facility. Fungi causing a "thread mold" defect in Cheddar cheeses were found in the cheese factory environment, on cheese-making equipment, in air, and in curd and whey.

In a study of cheese-making facilities in Denmark, *Penicillium commune* persisted in the cheese coating and unpacking areas over a 7-year period. Ascospores of *B. nivea* and other heat-resistant species shown to be able to survive pasteurization, such as *Talaromyces avellaneus, Neo-sartorya fischeri* var. *spinosa,* and *Eupenicillium brefeldianum,* have also been found in raw milk. A major cause of failure of processing and packaging systems is the development of biofilms on equipment surfaces. These communities of microorganisms develop when nutrients and water remain on surfaces between times of cleaning and reuse. Bacteria in biofilms (sessile form) are more resistant to chemical sanitizers than are the same bacteria in suspension (planktonic form). Chemical sanitizers may be rendered ineffective by biofilms leaving viable bacteria to be dislodged into the milk product.

PREVENTION OF SPOILAGE IN CULTURED DAIRY PRODUCTS

Cultured products such as buttermilk and sour cream depend on a combination of lactic acid producers, the lactococci, and the leuconostocs to produce the desired flavour profile. Imbalance of the culture, improper temperature or ripening time, infection of the culture with bacteriophage, presence of inhibitors, and/or microbial contamination can lead to an unsatisfactory product. A buttery flavour note is produced by *Leuconostoc mesenteroides* subsp. *cremoris*. This bacterium converts acetaldehyde to diacetyl, thus reducing the "green" or yogurt-like flavour. A diacetyl to acetaldehyde ratio of 4:1 is desirable, whereas the green flavour is present when the ratio is 3:1 or less.

Proteolysis by the lactococci is necessary to afford growth of the *Leuconostoc* culture, and citrate is needed as substrate for diacetyl production Although cooking of the curd destroys virtually all bacteria capable of spoiling cottage cheese, washing and handling of the curd after cooking can introduce substantial numbers of spoilage microorganisms. It is desirable to acidify alkaline waters for washing cottage cheese curd to prevent solubilization of surfaces of the curd. However, more pseudomonads can be adsorbed onto cottage cheese curd from wash water when adjusted to pH5 (40–45 per cent) rather than adjusted to pH7 (20–30 per cent). Flushing packages of cottage cheese or sour cream with CO2 or N2 suppressed the growth of psychrotrophic bacteria, yeasts, and molds for up to 112 days, but a slight bitterness can occur in cottage cheese after 73 days of storage. Cheesemakers can use the addition of high numbers of lactic acid bacteria to raw milk during storage to reduce the rate of growth of psychrotrophic microbes. For fresh, raw milk, brined cheeses, gassing defects can be reduced by presalting the curd prior to brining and reducing the brine temperature to <12°C. Pasteurization will eliminate the risk from most psychrotrophic microbes, coliforms, leuconostocs, and many lactobacilli, so cheeses made from pasteurized milk have a low risk of gassiness produced by these microorganisms.

Most bacterial cells, including spores, can be removed from milk by centrifugation at about 9,000*g*. The process, known as bactofugation, removes about 3 per cent of the milk, called bactofugate. Kosikowski and Mistry invented and patented a process for recovering this bactofugate which is heated at 135°C for 3–4 s, then added back to the cheese milk. The process can reduce the population of butyric acid-producing spores by 98 per cent. Spore-forming bacterial growth and subsequent gas production in aged, ripened cheeses can be minimized with a salt to moisture content of ^{3}3.0 per cent.

Other potential inhibitors of butyric acid fermentation and gas production in cheese are the addition of nitrate, addition of lysozyme cold storage of cheese prior to ripening, direct salt addition to the cheese curd, addition of hydrogen peroxide, or use of starter cultures that form nisin or other antimicrobials. The most popular mold inhibitors used on cheeses are sorbates and natamycin. Sorbates tend to diffuse into the cheese, thereby modifying flavour and decreasing their concentration, whereas very little natamycin diffuses.

Electron beam irradiation, studied by Blank, Shamsuzzaman, and Sohal for mold decontamination of Cheddar cheese, can reduce initial populations of *Aspergillus ochraceus* and *Penicillium cyclopium* by 90 per cent with average doses of 0.21 and 0.42 kGy, respectively. Since nearly all mold spores are killed by pasteurization practices that limit recontamination and growth, although difficult, are vital in prevention of moldy cheeses.

Modified atmosphere packaging (MAP) of cheeses can retard or prevent the growth of molds, and optimum MAP conditions for different types of cheeses were described by Nielsen and Haasum. For processed cheeses containing no active lactic acid starter bacteria, low O2 and high CO2 atmospheres were optimum; for cheeses containing active starter cultures atmospheres containing low O2 and controlled CO2 using a permeable film provided the best results.

For mold-ripened cheeses requiring the activity of the fungi to maintain good quality, normal O2 and high, but controlled, CO2 atmospheres were best. In Italian soft cheeses such as Stracchino, vacuum packaging decreased the growth of yeasts, resulting in a shelf life extension of >28 days. Processing times and temperatures used in the manufacture of cream cheese and pasteurized process cheese are able to eliminate most spoilage microorganisms from these products. However, the benefit of the presence of competitive microflora is also lost. It is very important to limit the potential for recontamination, as products that do not contain antimycotics can readily support the growth of yeasts and molds. Sorbates can be added; however, their use in cream cheese is limited to amounts that will not affect the delicate flavour.

PREVENTION OF SPOILAGE IN OTHER DAIRY PRODUCTS

The high salt concentration in the serum-in-lipid emulsion of butter limits the growth of contaminating bacteria to the small amount of nutrients trapped within the droplets that contain the microbes. However, psychrotrophic bacteria can grow and produce lipases in refrigerated salted butter if the moisture and salt are not evenly distributed. When used in the bulk form, concentrated (condensed) milk must be kept refrigerated until used.

It can be preserved by addition of about 44 per cent sucrose and/or glucose to lower the water activity below that at which viable spores will germinate (*a*w 0.95). Lactose, which constitutes about 53 per cent of the non-fat milk solids,

contributes to the lowered water activity. When canned as evaporated milk or sweetened condensed milk, these products are commercially sterilized in the cans, and spoilage seldom occurs. Microbial growth and enzyme activity are prevented by freezing. Therefore, microbial degradation of frozen desserts occurs only in the ingredients used or in the mixes prior to freezing.

DAIRY PRODUCTS

The dairy industry has come a long way since the early 1900s, when it began developing techniques for judging dairy products to stimulate interest and education in dairy science. In the traditional methods that emerged, judging and grading dairy products normally involved one or two trained "experts" assigning quality scores on the appearance, flavour and texture of the products based on the presence or absence of predetermined defects.

These traditional dairy judging methods have several shortcomings: they can't predict consumer acceptance; their quality assessments are subjective; assigning quantitative scores is difficult; and they don't combine analytically oriented attribute ratings with affectively oriented quality scores. Descriptive sensory profiles of two Cheddar cheeses that received the same grade by traditional grading techniques. With seven of the 11 flavour attributes measured as being significantly different between the two cheeses, the flavour perception of the two samples is actually quite different. Using traditional methods of evaluation, however, these products with very different sensory characteristics but no defect will obtain the same quality score.

SENSORY INPUT

One thing in common to all sensory assessment methods is that they use humans as the measuring instrument. There are many kinds of sensory tests, the most widely used being difference tests, descriptive analysis and consumer acceptance testing. Difference tests include the triangle test, in which the panel attempts to detect which one of three samples is different from the other two, and duo-trio tests, in which the panel selects which one of two samples is different from a standard. Difference tests estimate the magnitude of sensory differences between samples, but one deficiency of these tests is that the nature of the differences is not defined.

In most cases, a combination of difference tests and descriptive sensory analysis is employed for problem-solving. Descriptive sensory analysis refers to a collection of techniques that seek to discriminate between a range of products based on their sensory characteristics and to determine a quantitative description of the sensory differences that can be identified, not just the defects. Unlike traditional quality judging methods, no judgement of "good" or "bad" is made because this is not the purpose of the evaluation.

The panel operates as a powerful instrument to identify and quantify sensory properties. Descriptive sensory analysis provides useful information for dairy research, product development and marketing. Several assessors rating samples for a number of sensory attributes is a simple example of sensory profiling. For example, bitterness may be rated on a five-point scale, with a rating of one indicating no bitterness and a rating of five meaning very bitter. External standards may help to define attributes and standardize the scale for each assessor. Developing and refining a vocabulary, or sensory lexicon, is an essential part of sensory profile work and is done in an objective manner.

QUANTITATIVE DESCRIPTIVE ANALYSIS

The first published descriptive sensory technique is the Flavour Profile Method (FPM) developed in the 1950s by Arthur D. Little Inc. Refinements and variations in FPM occurred in the 1970s with the development of Quantitative Descriptive Analysis (QDA) and the Spectrum™ method of descriptive analysis.

Today, descriptive analysis has gained wide acceptance as one of the most important tools for studying issues related to flavour, appearance and texture, as well as a way to guide product development efforts. For example, it has been used as an investigative sensory technique for studying conventionally pasteurized milk, ice cream and cheese. With descriptive analysis, selected panelists work together to identify key product attributes and appropriate intensity scales specific to the product under study. The panelists are then trained by the panel leader, a sensory professional rather than a member of the panel, to reliably identify and score product attributes.

During training, the panel generates the language to describe the product. Descriptive analysis results are subjected to statistical analysis and are then represented in a variety of graphical formats for interpretation. One useful statistical technique is Principal Component Analysis (PCA), a multivariate analysis method that shows groupings or clusters of similar sample types based on quantitative measurements. By applying PCA to descriptive analysis data, the set of dependent variables (*i.e.*, attributes) is reduced to a smaller set of underlying variables (called factors) based on patterns of correlation among the original variables. The factors (also called principal components) are linear combinations of the independent variables.

The resulting data can then be applied in many useful ways. A few examples include profiling specific product characteristics, comparing and contrasting similar products based on attributes important to consumers, and altering product characteristics with the goal of increasing market share for a given set of products.

FLAVOUR LEXICONS FOR DAIRY PRODUCTS

M.A. Drake and G.V. Civille have reviewed lexicon history, methods and applications. A flavour lexicon is a set of word descriptors that describe a product's flavour. While the panel generates its own list to describe the product array under study, a lexicon provides a source of possible terms with references and definitions for clarification. Development of a representative flavour lexicon requires several steps, including appropriate product frame-of- reference collection, language generation and designation of definitions and references, before a final descriptor list can be determined.

Once developed, flavour lexicons can be used to record and define product flavour, compare products and determine storage stability, as well as to study correlations of sensory data with consumer liking/acceptability and chemical flavour data. Good flavour lexicons should be both discriminating and descriptive.

CHEDDAR CHEESE AND POWDERED MILK LEXICONS

M. A. Drake, at the Southeast Dairy Foods Research Center, developed and validated a descriptive language for Cheddar cheese flavour. For the project, 240 representative cheese samples were collected. Fifteen individuals from industry, academia and government participated in roundtable discussions to generate descriptive flavour terms. A highly trained descriptive panel (n=11) refined the terms and identified references.

Identification of chemical references was conducted with the assistance of K. Cadwallader at the University of Illinois. Instrumental analyses (gas chomatography/mass spectrometry, or GC/MS) were conducted to identify many flavour compounds that were responsible for specific flavours and off-flavours in Cheddar cheese. Twenty-four Cheddar cheeses were then presented to the panel to validate the proposed lexicon. The panel differentiated the 24 Cheddar cheeses as determined by univariate and multivariate analysis of variance. Twenty-seven terms were identified to describe Cheddar flavour. Seventeen descriptive terms were observed in most Cheddar cheeses. Drake's standard sensory language for Cheddar cheese today is facilitating training and communication among different research groups.

The Cheddar cheese lexicon is helping cheese-makers and cheese users accurately and consistently characterize the flavour of their cheese products and improve quality issues by measuring and controlling the presence of compounds that have been associated with flavour defects. Following development of the Cheddar cheese lexicon, Drake developed a similar language to help characterize another food industry staple: dried dairy ingredients, including whey proteins and non-fat dry milk. Global production of non-fat dry milk tops 3.3 million tons and whey protein demand still outstrips production, which increases annually.

A sensory lexicon describing the flavour of these ingredients helps dairy processors maximize the quality of these ingredients and allows food technologists to identify the exact attributes or flavour notes these ingredients contribute to formulations. Drake said she was surprised by the number of descriptive terms that the panel uncovered for application to the dried dairy ingredients lexicon. The panel discovered 21 flavour terms that could be applied to milk powders. Examples included cooked/milky flavour, cake mix or vanillin, sweet and sour, earth and cereal.

Each of these flavours was linked to a key aroma compound, many of which were identified by Drake and Cadwallader with GC/MS. For example, lactones tend to lend a sweet, coconut like flavour, while various free fatty acids can simulate a waxy flavour. Many different factors contribute to flavour variability. The source of the powder, processing/packaging methods and materials, as well as storage time and conditions, are just a few.

SCOPE FOR DAIRY FARMING AND ITS NATIONAL IMPORTANCE

The total milk production in the country for the year 2001-02 was estimated at 84.6 million metric tonnes. At this production, the per capita availability was to be 226 grams per day against the minimum requirement of 250 grams per day as recommended by ICMR. Thus, there is a tremendous scope/potential for increasing the milk production. The population of breeding cows and buffaloes in milk over 3 years of age was 62.6 million and 42.4 million, respectively (1992 census). Central and State Governments are giving considerable financial assistance for creating infrastructure facilities for milk production. The nineth plan outlay on Animal Husbandry and Dairying was ' 2345 crores.

FINANCIAL ASSISTANCE AVAILABLE FROM BANKS/ NABARD FOR DAIRY FARMING

NABARD is an apex institution for all matters relating to policy, planning and operation in the field of agricultural credit. It serves as an apex refinancing agency for the institutions providing investment andproduction credit. It promotes development through formulation and appraisal of projects through a well organised Technical Services Department at the Head Office and Technical Cells at each of the Regional Offices. Loan from banks with refinance facility from NABARD is available for starting dairy farming. For obtaining bank loan, the farmers should apply to the nearest branch of a commercial or co-operative Bank in their area in the prescribed application form which is available in the branches of financing banks. The Technical Officer attached to or the Manager of the bank can help/give guidance to the farmers in preparing the project report to obtain bank loan.

For dairy schemes with very large outlays, detailed reports will have to be prepared. The items of finance would include capital asset items such as purchase of milch animals, construction of sheds, purchase of equipments etc. The feeding cost during the initial period of one/two months is capitalised and given as term loan. Facilities such as cost of land development, fencing, digging of well, commissioning of diesel engine/pumpset, electricity connections, essential servants' quarters, godown, transport vehicle, milk processing facilities etc. can be considered for loan. Cost of land is not considered for loan. However, if land is purchased for setting up a dairy farm, its cost can be treated as party's margin upto 10 per cent of the total cost of project.

SCHEME FORMULATION FOR BANK LOAN

A Scheme can be prepared by a beneficiary after consulting local technical persons of State animal husbandry department, DRDA, SLPP etc., dairy co-operative society/union/federation/commercial dairy farmers. If possible, the beneficiaries should also visit progressive dairy farmers and government/military/agricultural university dairy farm in the vicinity and discuss the profitability of dairy farming. A good practical training and experience in dairy farming will be highly desirable. The dairy co-operative societies established in the villages as a result of efforts by the Dairy Development Department of State Government and National Dairy Development Board would provide all supporting facilities particularly marketing of fluid milk.

Nearness of dairy farm to such a society, veterinary aid centre, artificial insemination centre should be ensured. There is a good demand for milk, if the dairy farm is located near urban centre. The scheme should include information on land, livestock markets, availability of water, feeds, fodders, veterinary aid, breeding facilities, marketing aspects, training facilities, experience of the farmer and the type of assistance available from State Government, dairy society/union/federation. The scheme should also include information on the number of and types of animals to be purchased, their breeds, production performance, cost and other relevant input and output costs with their description.

Based on this, the total cost of the project, margin money to be provided by the beneficiary, requirement of bank loan, estimated annual expenditure, income, profit and loss statement, repayment period, etc. can be worked out and shown in the Project report. A format developed for formulation of dairy development schemes.

SCRUTINY OF SCHEMES BY BANKS

The scheme so formulated should be submitted to the nearest branch of bank. The bank's officers can assist in preparation of the scheme for filling in the prescribed application form. The bank will then examine the scheme for its technical feasibility and economic viability.

Technical Feasibility—this would briefly include:

- Nearness of the selected area to veterinary, breeding and milk collection centre and the financing bank's branch.
- Availability of good quality animals in nearby livestock market. The distribution of important breeds of cattle and buffaloes. The reproductive and productive performance of cattle and buffalo breeds.

- Availability of training facilities.
- Availability of good grazing ground/lands.
- Green/dry fodder, concentrate feed, medicines etc.
- Availability of veterinary aid/breeding centres and milk marketing facilities near the scheme area.

Economic Viability—this would briefly include:

- *Unit Cost*: The average unit cost of dairy animals for some of the States.
- Input cost for feeds and fodders, veterinary aid, breeding of animals, insurance, labour and other overheads.
- Output costs *i.e.* sale price of milk, manure, gunny bags, male/ female calves, other miscellaneous items etc.
- Income-expenditure statement and annual gross surplus.
- Cash flow analysis.
- Repayment schedule (*i.e.* repayment of principal loan amount and interest).

Other documents such as loan application forms, security aspects, margin money requirements etc. are also examined. A field visit to the scheme area is undertaken for conducting a techno-economic feasibility study for appraisal of the scheme. Model economics for a two animal unit and mini dairy unit with ten buffaloes.

DAIRY POLICIES TO ASSIST PRODUCERS

Indian dairy policy has been developed over the last seven decades.

The early policies addressed three main problems:

1. Producers lacked bargaining power with milk buyers;
2. Producers suffered from volatile or low prices; and
3. Market participants encountered severe shortages/gluts resulting from marketing a highly perishable commodity.

The policy response resulted in the development of two major government activities that still function today: federal milk marketing orders and the Dairy Product Price Support Programme. While both FMMOs and the DPPSP have their roots in the 1930s and 1940s, the programmes have changed modestly over the years as the industry structure and markets changed. Two other components of Indian dairy policy are relatively new programmes. First, the 1985 farm bill established the Dairy Export Incentive Programme to counter foreign competitor subsidies.

Second, the Milk Income Loss Contract programme was established in the 2002 farm bill as a government payment for dairy farmers in times of low milk prices. Like Indian crop programmes, the MILC programme pays dairy producers when prices decline below a specified level. The following sections describe each of these four components and how they relate to the current market situation. Lower milk and dairy product prices since late 2008 have generated new programme activity. Purchasing dairy products last fall under the DPPSP; MILC payments were triggered beginning in February.

MILK INCOME LOSS CONTRACT PROGRAMME

The Milk Income Loss Contract (MILC) programme pays dairy farmers when farm milk prices fall below an established target price. Section 1506 of the 2008 farm bill extends authority for the MILC programme until September 30, 2012. This programme is similar to long-time subsidy programmes for crops that pay farmers when farm prices drop below certain levels. Farm Service Agency implements the MILC programme. Under MILC, participating dairy farmers nationwide are eligible for a federal payment whenever the minimum monthly market price for farm milk used for fluid consumption in Boston falls below $16.94 per cwt.

Eligible farmers then receive a payment equal to 45 per cent of the difference between the $16.94 target price and the lower monthly market price. The payment quantity is limited to 2.985 million pounds of annual production.

Since the inception of the MILC programme, large dairy farm operators have expressed concern that the payment limit has negatively affected their income. For larger farm operations, their annual production is well in excess of the limit, and any production in excess of that receives no federal payments.

To address the issue of rising feed costs, the 2008 farm bill includes a provision that adjusts upward the $16.94 target price in any month when feed prices are above a certain threshold. The law requires calculating monthly a National Average Dairy Feed Ration Cost based on a formula that currently uses to calculate feed costs. In any month that the average feed cost is above

$7.35 per cwt., the $16.94 target price will be increased by 45 per cent of the difference between the monthly feed cost and $7.35. For the latter half of 2007 and all of 2008, farm milk prices remained well above the MILC trigger price, precluding the need for any MILC payments.

However, milk prices have since declined below the trigger for MILC payments. The Class I Boston farm milk price for February 2009 was $13.97 per cwt. With the adjustment for feed costs raising the trigger to $17.33 per cwt., MILC payments were activated for the first time in two years at a payment rate of $1.51. The payment rate rose to $2.01 per cwt. in March. Given current prospects in the futures markets for milk, corn, and soybeans, payments are expected to continue during 2009, but at smaller rates.

Individual producers must select which month to begin receiving payments, based on their projection of potential payment rates and the possibility of hitting the production payment limit. As of October 26, 2009, total MILC payments distributed to date were $775 million. The timing of the payments has caused some concern for producers this spring. While milk price data become available during the payment month, data needed for the feed cost adjustor are not available until publishes monthly average feed prices in *Agricultural Prices* at the end of the next month. Consequently, MILC payments for a particular month are not processed until two months later.

DAIRY PRODUCT PRICE SUPPORT PROGRAMME (DPPSP)

The Agricultural Act of 1949 first established a dairy price support programme by permanently requiring supporting the farm price of milk. Since 1949, Congress has regularly amended the programme, usually in the context of multiyear omnibus farm acts and budget reconciliation acts. Historically, the supported farm price for milk is intended to protect farmers from price declines that might force them out of business and to protect consumers from seasonal imbalances of supply and demand.

Commodity Credit Corporation (CCC) supports milk prices by its standing offer to purchase surplus non-fat dry milk, cheese, and butter from dairy processors. Whenever market prices fall to product support levels, processors generally make the business decision of selling surplus product to the government rather than to the marketplace. Consequently, the government purchase prices usually serve as a floor for the market price, which in turn indirectly supports the farm price of milk for all dairy farmers.

The effectiveness of the dairy price supports depends on removal of products from the market and placement into government storage. The Dairy Product Price Support Programme (DPPSP) as authorized by the 2008 farm bill requires purchasing products at the following minimum prices: block cheese, $1.13/lb.; barrel cheese, $1.10/lb.; butter, $1.05/lb.; and non-fat dry milk, $0.80/lb. Under previous law, the support price for farm milk was statutorily set at $9.90 per cwt., and given the administrative authority to establish a combination of dairy product purchase prices that indirectly supported the farm price of milk at $9.90. Although the 2008 law does not specifically state that the overall support price is $9.90 per cwt, each of the mandated product prices in the law is equivalent to the existing product purchase prices, so farm milk prices effectively continue to be supported at $9.90.

In late 2008 and 2009, after several years of relative inactivity, the price support programme resumed purchases when dairy product prices approached support levels. As of September 11, 2009, estimated that it purchased 111 million pounds of non-fat dry milk under the programme in 2008 and expects to purchase 379 million pounds in 2009, along with small amounts of butter and cheese (including amounts exported under the Dairy Export Incentive Programme).

Total expenditures on the DPPSP were $223 million from October 1, 2008, through September 10, 2009. With an expected rise in milk and product prices next year, Forecasts only a small amount of butter to be purchased in 2010. Following heightened industry and congressional interest in taking action to boost milk prices for farmers, July 31, 2009, a temporary increase in price support for cheese and non-fat dry milk from August 2009 through October 2009.

Subsequently, the Senate approved an amendment to the Senate-passed FY2010 agriculture appropriations bill to increase Farm Service Agency funding by $350 million, ostensibly for an additional increase in dairy product price support levels. However, the conference agreement for the FY2010 Agriculture appropriations bill, which was enacted on October 21, 2009, provides for a different use of the funds.

MILK MARKETING ORDERS

Federal milk marketing orders (FMMOs) mandate minimum prices that processors must pay producers for milk depending on its end use. This compares with the MILC programme, which provides direct payments to producers, and the DPPSP, which buys surplus dairy products at specified minimum prices. The DPPSP serves as a price floor for products and under girds FMMO minimum milk prices. The farm price of approximately two-thirds of the nation's fluid milk is regulated under FMMOs.

Federal orders, which are administered Agricultural Marketing Service, were instituted in the 1930s to promote orderly marketing conditions by, among other things, applying a uniform system of classified pricing throughout the market. Some states, California for example, have their own state milk marketing regulations instead of federal rules. FMMOs also address how market proceeds are distributed among producers delivering milk to federal marketing order areas.

Producers are affected by two fundamental marketing order provisions: the classified pricing of milk according to its end use, and the pooling of receipts to pay all farmers a blend price. Federal orders regulate dairy handlers (processors) who sell milk or milk products within a defined marketing area by requiring them to pay not less than established minimum class prices for the Grade A milk they purchase from dairy producers, depending on how the milk is used.

This classified pricing system requires handlers to pay a higher price for milk used for fluid consumption than for milk used in manufactured dairy products such as yogurt, ice cream, and sour cream, cheese and butter and dry milk products. These differences between classes reflect the different market values for the products. Blend pricing allows all dairy farmers who ship to the market to pool their milk receipts and then be paid a single price for all milk based on order-wide usage (a weighted average of the four usage classes).

Paying all farmers a single blend price is seen as an equitable way of sharing revenues for identical raw milk directed to both the higher-valued fluid market and the lowervalued manufacturing market. Manufactured class prices are the same in all orders nationwide and are calculated monthly based on current market conditions for manufactured dairy products. The Class I price for milk used for fluid consumption varies from area to area.

Class I prices are determined by adding, to a monthly base price, a "Class I differential" that generally rises with the geographical distance from milk surplus regions in the Upper Midwest, the Southwest, and the West. Class I differential pricing is a mechanism designed to ensure adequate supplies of milk for fluid use at consumption centers. The supply of milk may come from local supplies or distant supplies, whichever is more efficient.

However, local dairy farmers are protected by the minimum price rule against lower-priced milk that might otherwise be hauled into their region. Over the years, dairy farmers have supported minimum prices afforded by FMMOs because they help balance marketing power traditionally held by processors. In contrast, dairy processors generally oppose them.

Mandated minimum prices, they say, do not allow for timely adjustments in a rapidly changing market and can leave product manufacturers in unprofitable situations. Also, they contend that the FMMO system distorts markets, saying fixed differentials contributed to high fluid milk prices last year.

DAIRY EXPORT INCENTIVE PROGRAMME (DEIP)

First authorized in 1985, the Dairy Export Incentive Programme (DEIP) provides cash bonus payments to Indian dairy exporters. The programme was initially intended to counter foreign—mostly European Union—dairy subsidies (while removing surplus dairy products from the market), but subsequent farm bill reauthorizations have added market development to the role of DEIP. Payments since the program's inception have totaled $1.1 billion. The programme was active throughout the 1990s, peaking in 1993 with $162 million in bonuses.

DEIP funding is a mandatory account provided through the Commodity Credit Corporation (CCC) borrowing authority from the Indian Treasury, rather than through annual appropriations bills. The programme had not been used since FY2004 until announced its reactivation on May 22, 2009. Indian dairy product exports made with DEIP bonuses are subject to annual limitations under the Uruguay Round Agreement of the World Trade Organization (WTO).

The limits are 68,201 metric tons of skim milk powder, 21,097 tons of butterfat, 3,030 tons of various cheeses, and 34 tons of other dairy products (quantity limits are on a July-June year). Total expenditures under WTO commitments are now capped at $117 million per year (value limits on a October-September year).

MILK PRODUCTION LEVELS

A cow will produce large amounts of milk over her lifetime. Certain breeds produce more milk than others; however, different breeds produce within a range of around 15,000 to 25,000 lbs of milk per lactation. The average for dairy cows in the US in 2005 was 19,576 pounds. Production levels peak at around 40 to 60 days after calving. The cow is then bred. Production declines steadily afterwards, until, at about 305 days after calving, the cow is 'dried off', and milking ceases.

About sixty days later, one year after the birth of her previous calf, a cow will calve again. High production cows are more difficult to breed at a one year interval. Many farms take the view that 13 or even 14 month cycles are more appropriate for this type of cow. Dairy cows may continue to be economically productive for many lactations. Ten or more lactations are possible. The chances of problems arising which may lead to a cow being culled are however, high; the average herd life of US Holsteins is today fewer than 3 lactations. This requires more herd replacements to be reared or purchased.

Over 90 per cent of all cows are culled for 4 main reasons:

1. *Infertility*: Failure to conceive and reduced milk production. Cows are at their most fertile between 60 and 80 days after calving. Cows remaining

"open" (not with calf) after this period become increasingly difficult to breed, which may be due to poor health. Failure to expel the afterbirth from a previous pregnancy, luteal cysts, or metritis, an infection of the uterus, are common causes of infertility.

2. *Mastitis*: Persistent and potentially fatal mammary gland infection, leading to high somatic cell counts and loss of production. Mastitis is recognized by a reddening and swelling of the infected quarter of the udder and the presence of whitish clots or pus in the milk. Treatment is possible with long-acting antibiotics but milk from such cows is not marketable until drug residues have left the cow's system.

3. *Lameness*: Persistent foot infection or leg problems causing infertility and loss of production. High feed levels of highly digestible carbohydrate cause acidic conditions in the cow's rumen. This leads to laminitis and subsequent lameness, leaving the cow vulnerable to other foot infections and problems which may be exacerbated by standing in feces or water soaked areas.

4. *Production*: Some animals fail to produce economic levels of milk to justify their feed costs. Production below 12 to 15 litres of milk per day are not economically viable.

Herd life is strongly correlated with production levels. Lower production cows live longer than high production cows, but may be less profitable. Cows no longer wanted for milk production are sent to slaughter. Their meat is of relatively low value and is generally used for processed meat.

REPRODUCTION

Since the 1950s, artificial insemination (AI) is used at most dairy farms; these farms may keep no bull. Advantages of using AI include its low cost and ease compared to maintaining a bull, ability to select from a large number of bulls to match the anticipated market for the resulting calves, and predictable results.

More recently, embryo transfer has been used to enable the multiplication of progeny from elite cows. Such cows are given hormone treatments to produce multiple embryos. These are then 'flushed' from the cow's uterus. 7- 12 embryos are consequently removed from these donor cows and transferred into other cows who serve as surrogate mothers. The result will be between 3 and 6 calves instead of the normal single, or rarely, twins.

HORMONE USE

Hormone treatments are given to dairy cows to increase reproduction and to increase milk production. The hormones are used to produce multiple embryos have to be administered at specific times to dairy cattle to induce ovulation. Frequently, for economic considerations, these drugs are also used to synchronize a group of cows to ovulate simultaneously. The hormones Prostaglandin, Gonadotropin Releasing Hormone, and Progesterone are used for this purpose and sold under the brand names Lutalyse, Cystorelin, Estrumate, Factrel, Prostamate, Fertagyl. Insynch, and Ovacyst. They may be administered by injection, insertion or mixed with feed.

About 17 per cent of dairy cows in the United States are injected with Bovine somatotropin, also called recombinant bovine somatotropin (rBST), recombinant bovine growth hormone (rBGH), or artificial growth hormone. The use of this hormone increases milk production from 11 per cent-25 per cent, but also increases the likelihood of cattle developing mastitis, reduction in fertility and lameness. The U.S. Food and Drug Administration (FDA) has ruled that rBST is harmless to people, although critics point out increased levels of insulin-like growth factor 1 (IGF-1) in milk produced using this hormone. The use of rBST is banned in Canada, parts of the European Union, Australia and New Zealand.

NUTRITION

Nutrition plays an important role in keeping cattle healthy and strong. Implementing an adequate nutrition programme can also improve milk production and reproductive performance. Nutrient requirements may not be the same depending on the animal's age and stage of production.

Forages, which refer especially to hay or straw, are the most common type of feed used. Cereal grains, as the main contributors of starch to diets, are important in meeting the energy needs of dairy cattle. Barley is one example of grain that is extensively used around the world. Barley is grown in temperate to sub-artic climates, and it transported to those areas lacking the necessary amounts of grain. Although variations may occur, in general, barley is an excellent source of balanced amounts of protein, energy, and fibre.

Ensuring adequate body fat reserves is essential for cattle to produce milk and also to keep reproductive efficiency. However, if cattle get excessively fat or too thin, they run the risk of developing metabolic problems. Scientists have found that a variety of fat supplements can benefit conception rates of lactating dairy cows. Some of these different fats include oleic acids, found in canola oil, animal tallow, and yellow grease; palmitic acid found in granular fats and dry fats; and linolenic acids which are found in cottonseed, safflower, sunflower, and soybean. It is also important to note that proper levels of fat also improve cattle longevity.

Using by-products is one way of reducing the normally high feed costs. However, lack of knowledge of their nutritional and economic value limits their use. Although the reduction of costs may be significant, they have to be used carefully because animal may have negative reactions to radical changes in feeds. Such a change must then be made slowly and with the proper follow up.

PESTICIDE USE

A survey of the primary dairy producing areas in the US indicated that 13 per cent of lactating animals were treated with insecticides permethrin, pyrethrin, coumaphos, and dichlorvos primarily by daily or every-other-day coat sprays. Workers, particularly in stanchion barns, may be exposed to higher than recommended amounts of these pesticides.

BREEDS

In the United States, dairy cattle are divided into six major breeds. These are the: Holstein-Friesian, Brown Swiss, Guernsey, Ayrshire, Jersey, and Milking Shorthorn. In Rajasthan, an indigenous breed called Tharparkar exists, named from the Tharparkar District, now in Sindh Pakistan. Another type of dairy cow known as Nagauri from Nagaur District, the bull of which is renowned for its ability to plow fields and run. Traditionally, they used to pull covered wagons, known as rath, and in marriages to transport the newlywed couple. They are now a crutch for thriving agricultural and livestock rearing societies of the Thar Desert. Many other breeds are used nearly exclusively for beef, or for both dairy and beef purposes.

BUTTERMILK AND SOUR MILK PRODUCTS

Buttermilk is a by-product of the butter-making process. The taste can be more or less sour depending on the sourness of the cream or the milk, which is used for butter making, or on the degree to which it sours after churning. It is also possible to make a product like soured buttermilk using milk or skimmed milk, by inoculating it with sour milk and letting it ferment for one day. For the fermented milk you will need: fresh (skimmed) milk, a heat source, a wooden spoon, fresh fermented milk or buttermilk or a starter culture, a saucepan with a thick bottom, and a thermometer. Heat the fresh (skimmed) milk to boiling point, stirring all the time.

Cool it down to 18–20°C, for instance in a large pan with cold water. Add 10–30 ml of sour milk or buttermilk or a starter culture per each litre of milk (1 per cent). Leave for it 18-24 hours at room temperature (18–20°C);

if the surrounding temperature is higher, fermentation time will be somewhat shorter. After this the sour milk is ready. Store it in a cold place (cool basement or a refrigerator), if you want to keep it for some days.

RABI

Rabi is sweetened, concentrated milk. During concentrating, sugar is added from time to time.

You will need:

- (Unboiled) milk
- A heat source
- A wide, shallow iron pan with a thick bottom
- A flat metal scoop
- Sugar
- Scales.

Add sugar to the milk during the heating process (maximum 300 g per litre of milk) and follow the same procedure as for koa. Lumps of sugar will often be found in the end product.

YOGHURT

Yoghurt is produced when milk is soured by certain lactic acid bacteria, which prefer growing temperatures far above room temperature: 37–45°C. The milk should first be heated to 85°C or higher. A high pasteurisation temperature (above 72°C) gives a better consistency (thickness) to the final product. After the milk has been soured, the resulting yoghurt can be used to make more fresh yoghurt by adding it to fresh milk.

Basic Recipe for Yoghurt

You will need:

- Fresh raw milk
- A heat source
- A saucepan
- A spoon
- A thermometer
- Cooling facility (*e.g.* a large pan with cold water)
- A cool place (refrigerator or cellar)
- Starter culture for yoghurt or some fresh yoghurt
- Thermos flask or a box covered with a blanket

Heat the milk to 85°C or higher and keep it at this temperature for 3 minutes. Cool the milk to 45°C. Add 30 ml (2-3 tablespoons) of fresh yoghurt to each litre of milk; the yoghurt should not be more than 2 days old. Instead of fresh yoghurt you can use a yoghurt starter culture. Mix the milk and the starter and leave it to ferment. The time required for the milk to turn sour depends on the temperature.

To give you an idea:

- At 40-45°C it takes about 3 to 6 hours
- At 35-37°C it takes about 20 to 15 hours
- At 30°C it takes about 24 hours

The ideal temperature to make pleasant-tasting yoghurt with a firm consistency is 40-45°C. It is not possible to produce yoghurt at temperatures below 30°C or above 50°C. The correct temperature can be maintained using an insulated box or a blanket. Yoghurt is ready for consumption once the incubation period is finished. If cooled, yoghurt can be kept for one week.

Using a Thermos Flask

Heat the milk to at least 85°C, then cool it to 45°C. Pour 90 per cent of the milk into a thermos flask, which has been rinsed with hot water. Mix 1-2 tablespoons of fresh prepared yoghurt (or yoghurt culture) with the rest of the milk and add this to the thermos flask. Close the flask well and leave to stand for 3-6 hours. Remove the yoghurt from the thermos flask and store in a cool place. Yoghurt made of sheep milk is very firm and therefore not suitable for fermentation in a thermos flask.

Yoghurt made from Milk Powder

Make milk from milk powder according to the instructions on the package, but add 10 to 15 per cent extra milk powder. Dissolve the milk powder in water, heat it to boiling point and let it cool down to 45°C. Stir in 1-3 tablespoons of fresh yoghurt or a yoghurt culture per litre of milk. Cover the saucepan and put it in a warm, insulated place. After 3-6 hours this (firm and concentrated) yoghurt should be ready for consumption.

Remarks

- It is best to use fresh milk to make yoghurt. Milk powder can also be used. Sterilised milk may give a thinner yoghurt than pasteurised milk.
- After incubation, cooling is desirable, preferably below 10°C, so that souring is stopped (this retains the pleasant taste) and the bacteria

remain more viable, allowing the yoghurt to be used to inoculate milk again.

- Make sure the milk ferments as quickly as possible, preferably at 40-45°C rather than 30°C. Harmful bacteria have less opportunity to develop if the fermentation process goes faster.
- Thicker yoghurt can be produced by adding 2 to 3 tablespoons of milk powder to each litre of milk before heating it to 85°C.
- It is not advisable to use fruit yoghurt from a shop as a starter culture because it contains a lot of additives. Plain yoghurt from a shop can be used if it is not too old. Sterilised yoghurt is not suitable either, because the yoghurt bacteria have been killed by the sterilisation process.
- When using yoghurt from a carton or a pot as a starter, first remove the top layer and take the yoghurt from the centre to make fresh yoghurt. This is because the bacteria in the middle are probably the most diverse and active.
- Stir the product as little as possible before removing some, to avoid the extra risk of incorporating undesirable bacteria.

GHEE

To make ghee, you will need:

- Butter
- A heat source
- A pan
- A metal spoon.

Heat the butter until water and fat form separate layers; the fat will float on top.

There are two ways to remove the water:

1. It can be removed by further heating. The water present will evaporate
2. It is possible to remove the layer of fat with a spoon. This fat should then be heated again. The scum, which will form, has to be skimmed off regularly, preferably with a skimmer. The colour of ghee can vary from almost white to dark brown. A rancid flavour is acceptable, but if it tastes burnt it should be discarded.

8

Food Safety: Systems and Quality

FOOD SAFETY

INTRODUCTION

Microorganisms are tiny, mostly one-celled organisms capable of rapid reproduction under proper growth conditions. Those microorganisms important in the food industry include the *bacteria, viruses, yeasts, molds,* and *protozoans.* Many are helpful and serve useful functions such as causing breads to rise, fermenting sugars to alcohol, assisting in the production of cheese from milk, and decaying organic matter to replenish nutrients in the soil. Microorganisms can also cause foods to spoil and make them inedible. Spoilage organisms cost the food industry millions of dollars each year. Microorganisms can also be harmful.

These are called *pathogens* and cause between 24 to 81 million cases of foodborne illness in the U.S. each year. These forms of life, some so small that 25,000 of them placed end to end would not span one inch, were little known until the last century. Antony van Leeuwenhoek and others discovered "very little animalcules' in rain water viewed through crude microscopes. We now know that microorganisms occur everywhere on the skin, in the air, in the soil, and on nearly all objects. It was not until Pasteur proved that microorganisms could be eliminated from a system, such as a can of food, and sealed out (*hermetically sealed*), that man could exert control over the microbes in his environment.

Terminology

Bacteria are single-celled microorganisms found in nearly all natural environments. Outward appearances of the cell such as size, shape, and arrangement are referred to as *morphology.* Morphological types are grouped into the general categories of spherical (the cocci), cylindrical (the rods) and spiral. The cocci may be further grouped by their tendencies to cluster. Diplococci attach in pairs, streptococci in chains, staphylococci bunch like grapes, and sarcinae produce a cuboidal arrangement.

Bacterial cells have definite characteristic structures such as the *cell wall, cytoplasm,* and nuclear structures. Some also possess hairlike appendages for mobility called *flagella, fimbriae* which aid in attachment, plus *cytoplasmic* and *membranous inclusions* for regulating life processes. *Viruses* are extremely small parasites. They require living cells of plants, animals, or bacteria for growth. The virus is mainly a packet of genetic material which must be reproduced by the host. *Yeast and mold* are *fungi* which do not contain chlorophyls. They range in size from single-celled organisms to large mushrooms. Although some are multi celled, they are not differentiated into roots, stems and leaves. The true fungi produce masses of filamentous *hyphae* which form the *mycelium.* Depending on the organism, they may reproduce by fission, by budding as in the case of yeasts, or by means of *spores* borne on fruiting structures depending on the organism.

Protozoa are single-celled organisms such as the amoeba which can cause disease in humans and animals. They possess cell structure similar to higher, more complex organisms. Microorganisms are referred to by their scientific names which are often very descriptive. The first part of the name, the genus, is capitalized such as *Streptococcus,* spherical cells which occur in strips, *Lactobacillus* which are rod-shaped organisms commonly found in milk, or *Pediococcus* spherical cells which ferment pickles. The second part of the name is not capitalized and gives added information. Both parts of the name are underlined or italicized as in the case of *Saccharomyces cerevisiae,* a yeast which commonly ferments sausage.

The Cell

The cell is the basic unit of life. Our bodies are made up of millions of cells, but many microorganisms are single celled creatures. Cells are basically packages of living matter surrounded by membranes or walls. Within the cell are various organelles which control life processes for the cell such as intake of nutrients, production of energy, discharge of waste materials, and reproduction. Growth of the cell normally means reproduction. Bacteria and similar organisms reproduce by *binary fission*, a splitting of a single cell into two. The control center for the bacterial cell is the *nuclear structure.*

Within it is the genetic material which is duplicated and transferred to daughter cells during reproduction. These daughter cells can again divide to produce four cells from the original one. The time It takes for a new cell to produce a new generation of daughter cells is called *generation time.* Under optimum growth conditions, certain organisms can have a generation time of 15 minutes. In four hours over 65,000 cells could be produced from a single microorganism! Under adverse conditions, certain bacteria can protect the cell's genetic material by producing *spores*. These are extremely resistant capsules of genetic materials. Though there are no discernible life processes in the spore, under proper sporulation conditions, a viable, reproducing cell will germinate from it.

Factors Affecting Growth

Microorganisms, like other living organisms, are dependent on their environment to provide for their basic needs. Adverse conditions can alter their growth rate or kill them.

Growth of microorganisms can be manipulated by controlling:

- Nutrients available
- Oxygen
- Water
- Temperature
- Acidity and pH
- Light
- Chemicals

Nutrients

Nutrients such as carbohydrates, fats, proteins, vitamins, minerals and water, required by, man are also needed by microorganisms to grow. Microbes differ in their abilities to use *substrates* as nutrient sources. Their enzyme systems are made available according to their genetic code. They vary in ability to use nitrogen sources to produce amino acids and, therefore, proteins. Some require amino acids to be supplied by the substrate. When organisms need special materials provided by their environment, we refer to them as *fastidious.* Difference in the utilization of nutrients and the waste products they produce are important in differentiating between organisms.

Oxygen

Microbes also differ in their needs for free oxygen. *Aerobic* organisms must grow in the presence of free oxygen and *anaerobic* organisms must grow in the absence of free oxygen. *Facultative* organisms can grow with or without oxygen, while *microaerophilic* organisms grow in the presence of small quantities of oxygen.

Water

Water is necessary for microbes to grow, but microbes cannot grow in pure water. Some water is not available. A measurement of the availability of water is aw or *water activity.* The aw of pure water is 1.0 while that of a saturated salt solution is 0.75. Most spoilage bacteria require a minimum aw of 0.90.

Some bacteria can tolerate an aw above 0.75 as can some yeasts and most molds. Most yeasts require 0.87 water activity. An aw of 0.85 or less suppresses the growth of organisms of public health significance.

Temperature

Microorganisms can grow in a wide range of temperatures. Since they depend on water as a solvent for nutrients, frozen water or boiling water inhibits their growth. General terms are applied to organisms based on their growth at different temperatures. Most organisms grow best at or near room and body temperature. These are *mesophiles.* Those growing above 400C (1050F) are called *thermophiles* while those growing below 250C(750F) are called psychrotrophs.

Acidity

The nature of a solution based on its acidity or alkalinity is described as pH. The pH scale ranges from 0, strongly *acidic,* to 14, strongly *basic.* *Neutral* solutions are pH 7, the pH of pure water. Most bacteria require near neutral conditions for optimal growth with minimums and maximums between 4 and 9. Many organisms change the pH of their substrate by producing by-products during growth. They can change conditions such that the environment can no longer support their growth. Yeasts and molds are more tolerant of lower pH than the bacteria and may outgrow them under those conditions.

Light & Chemicals

Ultraviolet light and the presence of chemical inhibitors may also affect the growth of organisms. Many treatments such as hydrogen peroxide and chlorine can kill or injure microbes. Under certain conditions those given a sublethal treatment are injured, but can recover.

Growth

Characteristic growth patterns can be illustrated on a graph. There is a selected portion of the normal growth curve which is referred to as the *logarithmic growth phase* or the *log phase.* When cells begin to grow, we usually observe a period of no apparent growth which we refer to as the *lag phase.* This occurs because cells are making necessary adjustments to adapt. Next we experience the rapid growth or the *log phase* previously described. As cell mass becomes large, nutrients are exhausted and metabolic byproducts collect. Growth tapers off and the population remains constant for a time. This is referred to as the *stationary phase* of growth. With no intervention in the system the population will enter a *death phase* and total numbers of organisms will decline.

Enumeration of Cells

Numbers of microorganisms can be estimated based on cell counts, cell mass, or activity. A *direct count* of cells may be made by examination of a known volume of cell suspension under a microscope. This method is rapid and requires minimal equipment. This does not distinguish living cells from dead and may be tedious.

The application of certain stains make visible morphological characteristics of the organism which can aid in identification. Probably the most common method of enumerating cells is the *plate count.* A known volume of a diluted specimen is added to agar in a petri dish. Assuming dilute solution and that each organism will divide until it develops a visible mass or *colony,* the colonies can be counted and multiplied by the dilution factor to estimate the number of organisms in the original sample.

This method is based on the assumption that a colony forming unit (CFU) is a single organism. This will not hold exactly true in the case of strips or clumps of cells. Sublethally injured organisms may not grow. The culture conditions may not be conducive to the growth of certain types of *fastidious organisms* such as anaerobes. To measure the progress of a culture in a clear broth, changes in *turbidity* can be measured and related to numbers or organism. This method is easy and rapid. Many commercial establishments have found application for this method.

Identification

As many types of cells look similar in morphology and produce similar colonies, it becomes necessary to identify the organisms by their biochemical characteristics. Biochemical testing requires pure cultures isolated from a single colony from a plate count or streaked plate made for isolation purposes. *Isolates* are grown in an enriched broth to produce large cell numbers. Various media can then be inoculated with the culture and then growth can be observed by carefully formulating the various media, the biochemical and growth characteristics of the organism can be determined. Previously determined morphological characteristics can be combined with biochemical data to properly classify the organism. Newer methods more rapidly identify organisms of interest using other characterization such as monoclonal antibodies and DNA.

Thermal Processing of Foods

Low-acid canned foods are regulated by 21CFR113. These foods have a pH of greater than 4.6 and have a aw greater than 0.85. the regulations require that a *scheduled process* established by a processing authority be selected by the manufacturer which renders the product, under the specified conditions, *commercially sterile.* Commercial sterility is determined by processing food inoculated with known quantities of microbial spores. The test organisms used should simulate the resistance of *Clostridium botulinum* under those conditions. A process which eliminates these spores will destroy *Clostridium botulinum*

spores. Acid and acidified foods will not allow the growth of *Clostridium botulinum.* However, a *pasteurizing* heat treatment is necessary to destroy other bacteria, viruses, yeasts, and mold. Temperatures and process times which destroy microorganisms without destruction of nutrients may fail to deactivate enzymes. If the process selected does not inactivate the enzymes, product changes may proceed during storage, at an accelerated rate, and cause a loss of product quality. The microbiological quality of raw product is the simple greatest determinant in the level of quality in the food. Thermal processing is no substitute for good raw product quality.

The Retail Food Safety Need

Introduction

As more retail food operations across the U.S. and throughout the world compete to feed consumers, it becomes essential that uniform hazard analysis and control guidelines for producing, buying, and selling food products be developed. These guidelines must be based on science and validated in actual operation. At this time, consumers in the U.S. are doing less food preparation themselves and are dining out and/or are relying on retail food outlets for ready-prepared items. Food operations, as defined in this document, include: food markets where food is sold to be prepared in the home; food preparation and foodservice establishments that include restaurants, institutional foodservice units, street vending operations, hotel and lodging operations, military commissaries; and even the home, which is actually a miniature foodservice unit.

Food science and technology have improved the understanding of the potential microbiological, chemical, and physical hazards in foods. This knowledge can be used to determine the criteria necessary to assure that food products and commodities meet consumer safety expectations with an acceptable risk at the raw material level, the distributor level, and the consumer level. International trade and tourism will be enhanced throughout the world when there is a clearer understanding between the producer, retailer/supplier, and the buyer of food concerning the potential hazards in food and the level of risk associated with consuming a food.

Beginning with *Codex Alimentarius* (32) and the International Commission on Microbiological Specifications for Food (ICMSF), and continuing with the National Advisory Committee on Microbiological Criteria for Foods (135), there has been a movement for many years for more complete safety specifications for foods in local, national and international trade. The result is the current emphasis on Hazard Analysis and Critical Control Points (HACCP) in food production facilities and retail food operations. However, people have lost sight of the fact that HACCP is only a part of a company's food production quality management program. A company cannot accomplish process hazard control until it has process quality control. Hazards and critical control points can be easily identified. However, it is a separate issue to actually operate so that there is a very low chance of process deviation and low risk of a hazardous item being produced.

Hazard Classification

The World Health Organization (WHO) Division of Food and Nutrition (200) identifies a hazard as a biological, chemical, or physical agent or condition in food with the potential to cause harm. In addition to these factors, nutrient levels inadequate to prevent deficiency disease must also be considered as a hazard. Note that one food, or even a single serving, while it could cause foodborne illness, no retail dish or retailer (except in institutions) is responsible for balancing the client's entire diet.

Actually, food is never risk free. The Environmental Protection Agency (EPA) has developed many risk criteria for water and chemicals. Unfortunately, the consumer's standard for purchased food is zero risk, which is unattainable. There will never be totally pathogen-free food animals or fruits and vegetables, nor will there ever be human populations who excrete totally pathogen-free fecal material. Therefore, hazards must be assessed, and risks must be reduced to a safe level.

Hazard analysis is the process of collecting and interpreting information to assess the risk and severity of potential hazards (ICMSF, 90). It is appropriate to classify hazards as critical concern, major concern, and minor concern.

The definitions for hazards in food are as follows:

- *Critical Concern*: Without control, there is life-threatening risk.
- *Major Concern*: A threat that must be controlled but is not life threatening and requires no government intervention.
- *Minor Concern*: No threat to the consumer (normally quality and cleanliness issues).

For example, spoilage bacteria, even at more than 50,000,000 per gram, are of no known safety concern. Coliform bacteria include both spoilage and pathogenic microorganisms. Therefore, a coliform count of 1,000 CFU per gram is of minor concern until specified levels of specific pathogens in the coliform group are established. Only pathogens and pathogenic substances ingested above threshold levels can cause illness, disease, and death. Properly controlled levels of salt, sugar, and MSG are of no concern. Floors, walls, ceilings, and many other items grouped under Good Manufacturing Practices (31) are really minor concern. Of minor concern is also the presence of 1,000 *Staphylococcus aureus* cells, *Bacillus cereus* spores, or *Clostridium perfringens* spores per gram of food. These organisms are not hazardous until they reach 100,000 vegetative cells per gram.

A hazard of critical concern, on the other hand, is a dose of 1 or more *Escherichia coli* O157:H7, 100 *Salmonella* spp., or 500 *Campylobacter jejuni* in a portion of hamburger or chicken. It is essential that food be prepared by a cook who is trained to reduce potential hazards to a safe level. Even healthy people can become ill if they consume food containing these pathogens at high enough levels.

Some people develop natural immunity. People who live in an environment with greater levels of pathogenic agents have a greater tolerance. The example is farmers who acquire a natural immunity and elevated resistance to some illnesses after being exposed to many of these pathogens while working on their farms. It is the job of the cook to make food safe by washing the food, such as raw fruits and vegetables and by pasteurizing raw food with the application of heat. During preparation, care must be taken to prevent cross-contamination of any ready-to-eat foods when raw foods are handled. (Raw foods include fruits and vegetables as well as meat, fish, and poultry products.)

The Process of Hazard Identification

In order to develop a process or operation capable of protecting public health, which, in turn, will minimize liability costs, it is essential that a logical process for hazard identification be followed.

The following criteria should be included in the hazard analysis:

1. Evidence that the microbial, chemical, or physical agent is a hazard to health based on epidemiological data or a laboratory hazard analysis.
2. The type and kind of the natural and commonly acquired microflora of the ingredient or food, and the ability of the food to support microbial growth.
3. The effect of processing on the microflora of the food.
4. The potential for microbial contamination (or recontamination) and/or growth during processing, handling, storage, and distribution.
5. The types of consumers at risk.
6. The state in which food is distributed (e.g., frozen, refrigerated, heat processed, etc.).
7. Potential for abuse at the consumer level.
8. The existence of Good Manufacturing Practices (GMPs).
9. The manner in which the food is prepared for ultimate consumption (i.e., heated or not).
10. Reliability of methods available to detect and/or quantify the microorganism(s) and toxin(s) of concern.
11. The costs/benefits associated with the application of items 1 through 10 (as listed above).

Coupling hazard analysis with correct hazard control and operating procedures enables the retail food operator to demonstrate a high degree of "due diligence" in the prevention of problems. Due diligence is essential if an operator is to avoid punitive legal damages resulting from a lawsuit because he/she did nothing to control the hazards in the food.

Food-related Illness and Death in the United States

While cooks can control most pathogenic agents in food most of the time, there are still an enormous number of illnesses and deaths that occur each year in the U. S because of foodborne agents. More than 200 known diseases are transmitted through food. The causes of foodborne illness and disease include viruses, bacteria, parasites, toxins, metals, and prions, and the symptoms of foodborne illness vary from mild gastroenteritis to life-threatening neurologic, hepatic, and renal disorders.

A 1999 Centers for Disease Control and Prevention report (203) estimated cases of illness and deaths for the general population in the U.S. due to known causes. It can be seen that not all of these pasthogens are transferred by food. They are sometimes spread by water, person-to-person contact, or other means. Also note that 67% of the estimated 13.8 million foodborne illnesses of known etiology are viral (Norwalk-like viruses).

The authors of the 1999 CDC report (203) applied another analysis, taking into account under-reporting factors, to arrive at an estimate that predicts a much larger annual illness incidence of 76 million (mostly diarrheal illnesses of 1-to-2-day duration), 325,000 hospitalizations, and 5,000 deaths. This estimate is based on the authors' further speculation that 80% of the illnesses that occur annually are due to "unidentified etiological agents". The CDC report (203) did not estimate causes of illness or injury due to chemicals, toxins, or hard foreign objects in food. Predicted annual occurrences cited in Table below are those estimated by Todd (186).

The pathogens responsible for foodborne illness and disease are ubiquitous. Because of environmental and animal contamination, food and food products will always be contaminated with low levels of pathogens, and occasionally at high levels (especially poultry). At low levels, pathogenic microorganisms in food cause no problems, and people even develop immunity to their presence in food (74, 85, 91, 137, 151, 181, 198). At illness thresholds, however, pathogens in food can make people ill and cause death. Pathogens in food can only be controlled when food producers, food retailers, and consumers know the potential hazards in food, and handle and prepare food by methods to reduce the hazard to a tolerable level.

How do pathogens in food get to high levels? It can happen anywhere in the food chain, from growing the animal, fish, vegetable or fruit to consumption. Everyone in the food chain must have knowledge of the causes of foodborne disease and illness, and must establish a program that assures safety before they produce and sell food. If not, incomplete hazard control processes are implemented.

Government Microbiological Standards for Raw and Pasteurized Food The Code of Federal Regulations has been interpreted by the United States Department of Agriculture and the Food and Drug Administration to mean that if a sample of a processed food is found to be contaminated with *E. coli* O157:H7, *Salmonella* spp., or *Listeria monocytogenes*, the food is deemed unfit for human consumption. Sample size is variable, often negative in 25 grams, but the products' microbial standard is not enforced in retail operations, because retail inspectors seldom check. In retail food operations, the presence of *L. monocytogenes* is likely on fresh produce, meat, fish, and poultry, as well as on floors and in floor drains. It has been estimated by Farber that raw food (e.g., coleslaw) can contain 100 to 10,000 CFU/g of *L. monocytogenes*. Meat products will have low levels of *E. coli* O157:H7 and *Salmonella* spp., and poultry is often found to be contaminated with *Campylobacter jejuni* and *Salmonella* spp.

Need for International Safe and Hazardous Level Guidelines: At the present time, there are few worldwide food safety guidelines for upper or lower control limits of potential hazards in foods, or the process limits. This document proposes limits. When standards do exist, they may be inappropriate (e.g., specification of the numbers of coliforms in milk and in shellfish waters), or they may be unattainable (e.g., a zero level of both *Salmonella* spp. and *L. monocytogenes* in food). There is no zero level in food safety. There is a point at which measurements cannot be made with any degree of statistical reliability, and this point is frequently taken as "zero." For example, processed food is assumed to be safe from salmonellae contamination if there are no detectable salmonellae in a sample using the method of analysis described by the Bacteriological Analytical Manual (54). However, as laboratory methods improve, standards for safe levels of pathogenic material in food may be established. When safety standards or guidelines are developed for microbiological, chemical, and physical hazards in food, the standards or guidelines must be based on the risk of causing injury or illness to consumers, not what the processing industry is capable of achieving, or what scientific technology is capable of measuring.

Contamination Levels and Microbiological Control

Pathogen Contamination from Human Sources

In addition to the contamination of raw food supplies that occurs during growing, shipping, and processing, there is the problem of food contamination caused by people who are carriers of pathogens. While most food codes require that when an employee is sick, he or she should stay home, people actually shed pathogenic bacteria a few hours to many days before they have major symptoms of illness. Food workers can become permanent carriers of pathogens and yet exhibit no signs of illness (e.g., *Salmonella* carriers). Therefore, the only safe assumption is that all employees who work with food every day carry pathogens (on their skin and in their urine and feces) that must be kept out of the food. The following table is a list of some pathogens that commonly originate from human sources.

Table. Pathogen Contamination Common from Human Sources*

Microorganism	Source
Shigella spp., hepatitis A, Norwalk virus, *E. coli, Salmonella* spp., *Giardia lamblia*	Feces
Norwalk virus	Vomit
Staphylococcus aureus	Skin, nose, boils, skin infections
Streptococcus Group A	Throat and skin

Note:

** 1 in 50 (2%) of the employees who come to work each day are highly infective. (This calculation is based on 16,000,000 people who are ill each year for 2.5 days, of which 5,000,000 work in food operations each day.)*

Foodborne Illness Hazards: Threshold and Quality Levels

The next step in the systematic approach to hazard control is to recognize that certain pathogens can be selected as the basis for process control criteria. Table shows the levels of microorganisms that caused illness, based on volunteer feeding tests of healthy people. Hazardous levels of chemicals and hard foreign objects in food are also listed.

Spores of the pathogenic bacteria *B. cereus, C. botulinum,* and *C. perfringens* must germinate after the food is cooked, and the vegetative cells must multiply to a level of at least 104 to 106 in order to make the food a hazard to consumers. Low levels of these spore-forming organisms (102 to 103) in food are not a hazard, with the exception of *C. botulinum* found in honey. Honey must not be fed to infants at an age of less than one year (103). Infants have few competitive gut microflora, and the *C. botulinum* in the honey is able to germinate and then multiply in the infant's intestinal tract and produce toxin at a level sufficient to cause illness and even death. This hazard can be controlled by educating family members not to feed honey to babies.

Campylobacter jejuni is a principal cross-contamination problem. This pathogenic microorganism may be present in poultry at a "natural" level after slaughter that will cause illness without having to multiply. Raw chicken and other poultry can be contaminated at high levels (i.e., 106 to 107 organisms / chicken) (84). The threshold for illness in healthy people is approximately 500 organisms in 180 ml of milk, which means approximately 2 organisms per ml. Comparing this infective level with those for *Salmonella, E. coli, Vibrio,* etc., it is obvious that *C. jejuni,* in poultry that is grossly contaminated, will be a major cause of foodborne illness.

Other human fecal pathogens of concern include *Shigella* spp., pathogenic *E. coli, Salmonella* spp., hepatitis A virus, and Norwalk virus. For example, human fecal material from infected individuals can contain millions of shigellae per gram. Since toilet paper is unreliable in preventing fingertips from coming in contact with fecal material, fingertip washing becomes critical. Note that safe food for immune-compromised people in the U.S. has been basically defined by the government as no detectable *Salmonella* spp. or *L. monocytogenes* in one or two 25-gram samples from a lot. A level of less than 100 CFU *L. monocytogenes* per gram in ready-to-eat food, as set by Canada (46, 47), is probably realistic. Immune-compromised people must take the responsibility themselves for staying healthy. For example, they must understand that typical foodservice salads are high-risk food items for containing *L. monocytogenes* and should be avoided. These individuals should insist on well-cooked food.

There are very few sources of data to indicate the threshold levels for hepatitis A virus and Norwalk virus. If *Shigella* spp. is controlled, hepatitis A and Norwalk virus will probably also be controlled. Assuming toilet paper is 99.9% effective, 106 of the 109 pathogens in the feces of an ill employee will get on the employee's fingertips and underneath his or her fingernails. The fingertip washing process must reduce these organisms to below 10 in order to prevent an employee from transferring this pathogen to food that will make customers ill. Since antimicrobial chemicals used on hands only reduce pathogens about 100 to 1, the key safety strategy involves using a good hand soap or detergent; physical agitation of the fingertips with a fingernail brush; a lot of warm, flowing water; followed by a second hand wash without the brush. A greater than 105 reduction can be achieved using this double hand wash procedure.

Threshold levels for some chemical additives are also listed in Table below. There are hundreds of additives that are accepted for food use. The addition of these compounds to food must comply with level of use defined by the Code of Federal. It is essential that all chemical additives must be measured correctly before being added to food. For instance, in foodservice, there are no guidelines for the amount monosodium glutamate (MSG) in food. Hence, MSG is commonly overused in some restaurants. The 0.5% limit of use for monosodium glutamate suggested in Table below is based on 1/3 of the levels recommended by a noted supplier of MSG

Assumed Contamination Levels for Raw Food

A series of beginning contamination levels for raw food coming into typical foodservice systems is shown in Table. It is based on threshold levels that make healthy people ill and normal contamination as listed in the literature.

These figures pertain to food in the U.S. Contamination levels in other countries may be higher or lower. When designing a safe food process, these are the levels of contamination that must be controlled and eliminated by methods of processing, preservation, and storage.

Table 4-3. Expected Per Gram

Microorganisms	Meat & Poultry (CFU / g)
Salmonella spp., Vibrio spp., Hepatitis A, Shigella spp., E. coli, L. monocytogenes	10
Campylobacter jejuni	1,000*
Clostridium botulinum	0.01
Clostridium perfringens	100
Bacillus cereus	100
Mold toxins	Below governme
Chemicals & poisons	Below governme

* In poultry

Table 4-3. Expected Per Gram Pathogen Contamination on Raw Food

Microorganisms	Meat & Poultry (CFU / g)	Fish & Shellfish (CFU / g)	Fruits & Vegetables (CFU / g)	Starches (CFU / g)
Salmonella spp., *Vibrio* spp., Hepatitis A, *Shigella* spp., *E. coli*, *L. monocytogenes*	10	10	10	10
Campylobacter jejuni	1,000*			
Clostridium botulinum	0.01	0.01	0.01	0.01
Clostridium perfringens	100	10		
Bacillus cereus	100	10	100	100
Mold toxins	Below government tolerances			
Chemicals & poisons	Below government tolerances			

* In poultry

Often there is the question for retail food operators, "What is a microbiological standard for good food quality?" Neither the USDA nor the FDA provides an answer to this question. Microbial specifications that can be used when communicating with suppliers as to the microbial quality of raw food (171).

Table. Microbial Criteria for Food Quality

Number of Spoilage Microorganisms Aerobic Plate Count at 70°F (21.1°C)	Rating
< 10,000 CFU / gram	Good
10,000 to 5,000,000 CFU / gram	Average
5,000,000 to 50,000,000 CFU / gram	Poor
> 50,000,000 CFU / gram	Spoiled

The purpose of foodservice is to provide food that is safe and pleasurable to consume. Therefore, in addition to providing safe food by controlling and eliminating pathogens in food, the growth of spoilage microorganisms must be limited and controlled as much as possible in order to provide food products of a specified quality.

Assumed Microbiological Criteria for Food Handlers and Food Contact Surfaces In addition to microbial contamination expected in food, it is also necessary to define contamination levels for food handlers, facilities, and equipment. Based on the estimate that as many as 16 million people may get foodborne illness each year (15), for a period of 2 to 5 days. It can be estimated through calculations that 1 in 50 people working in the 500,000 food establishments in the U.S. is shedding billions of pathogens in his or her feces. If 0.001 gram of fecal material leaks through or gets around toilet paper, fingertips and underneath fingernails will become contaminated with fecal pathogens.

Hands and fingertips can also become contaminated when changing diapers, cleaning up vomit, or cleaning up after animals at home. The double hand wash method using a fingernail brush must be used to reduce fecal microorganisms to less than 10 on the fingertips and under the fingernails. The facilities and equipment will also be contaminated. If the food contact surfaces are washed and rinsed every 4 hours, and if the facilities are cleaned and sanitized adequately at the end of production, the pathogenic build-up on the equipment and facilities can be kept to a safe level. Note that *L. monocytogenes* is an environmental pathogen that arrives on food or on the people who enter the facility.

It is essential that the facility be as well maintained as possible in critical locations, so that cracks in the floors, walls, or ceilings (where *L. monocytogenes* can accumulate) are minimal. When cleaning is done regularly, *L. monocytogenes* can be reduced to an undetectable level immediately after cleaning. It is also essential that food processing areas be kept at a humidity of less than 65% to minimize mold multiplication in dry food and in the environment of the facility.

In a food production area where pasteurized, cooked, cooled food is being assembled and packaged for refrigerated meals, there must be no pathogens on food contact surfaces. This is defined as no vegetative pathogens (e.g., *Salmonella* spp., *Shigella* spp., and *C. jejuni*) in a 50-sq.-cm. swabbed area of a surface. The Standards for Number of Spoilage Microorganisms on Food Contact Surfaces are shown in Table. If these standards are maintained, industry experience has shown that foods with long shelf lives can be produced.

Table. Standards for Number of Spoilage Microorganisms on Food Contact Surfaces

Number of Spoilage Microorganisms	Rating
<1 CFU / 50 sq. cm. or <1/ml.	Excellent of rinse solution
2 to 10 CFU / 50 sq. cm.	Good
11 to 100 CFU / 50 sq. cm.	Clean-up time
101 to >1,000 CFU / 50 sq. cm.	Out of control, shut down and find the problem

Food Pathogen Control Data Summary

The next step is to develop the microbiological basis for the time and temperature and pH process standards. Table below provides the database for process standards development.

Since *Y. enterocolitica* and *L. monocytogenes* both begin to multiply at 29.3^{0}F (-1.5^{0}C) (76), food must be kept below a temperature of 30^{0}F (-1.1^{0}C) if the multiplication of these pathogens is to be totally stopped. This means only having frozen food, which is impractical. Another solution is to use time with temperature from 29.3 to 127.5^{0}F (-1.5 to 52.2^{0}C) so that the pathogens do not multiply to an unsafe level. *Salmonella* spp. will multiply at a pH as low as 4.1. Therefore, it is essential that if food such as mayonnaise is made with raw eggs (notorious for being contaminated with *Salmonella* spp.), the pH of the product must be below 4.1. Smittle (167) determined that there is no *Salmonella* spp. growth at pH 4.1 (when acidified with acetic acid), and in fact, *Salmonella* spp. in mayonnaise is destroyed when held at 70^{0}F (21.1^{0}C) for 72 hours.

The data for *C. jejuni* point out that it grows very poorly over a limited range and is quite easily destroyed. Therefore, the major problem with *C. jejuni* is cross-contamination, as mentioned earlier. It can be assumed that 10,000 *Campylobacter* spp. per 50 square cm (177) will be deposited on the food contact surface by raw food such as chicken, and it must be reduced to less than 100 per 50 square cm to be tolerable.

Spores of *C. botulinum* type E are inactivated at 180^{0}F (82.2^{0}C). However, many foods do not reach this temperature when cooked or pasteurized. It must be assumed, then, that *C. botulinum* type E and other non-proteolytic *C. botulinum* spores will survive the cooking process.

Many foods, 0.25 inch below the surface, have an oxidation reduction potential at which *C. botulinum* will grow. For control, if the food is a probable carrier of non-proteolytic *C. botulinum,* the food must be stored below 38^{0}F (3.3^{0}C) in order to prevent outgrowth of the spores. *Staphylococcus aureus* begins to multiply at 43^{0}F (6.1^{0}C) but does not begin to produce a toxin until it reaches a temperature of 50^{0}F (10.0^{0}C). Since there may be recontamination of food with *S. aureus* when people make salads with their hands, if salad ingredients are pre-chilled to less than 50^{0}F (10.0^{0}C) and are kept below this temperature when mixed, there will be no chance of *S. aureus* toxin production. Food containing 1,000 *S. aureus* organisms per gram is not hazardous. A population of 106*S. aureus* per gram of food is necessary to produce a sufficient amount of toxin to cause illness.

However, if the toxin is produced, it is virtually impossible to destroy the toxin when the food is cooked or reheated. Therefore, reheating food to 165^{0}F (73.9^{0}C) should never be used as a critical control procedure. After food is cooked, the best method of control is to prevent the production of toxin by controlling cross-contamination and by keeping the temperature of food below 50^{0}F (10.0^{0}C). Some types of *B. cereus* begin to multiply below 40ºF (4.4^{0}C) (192). It is a very common contaminant of many cereal products, spices, and even fresh sprouts. This spore-forming pathogen will survive cooking. To be safe for long-term storage (greater than 10 days), cooked food must be cooled and maintained at less than 38^{0}F (3.3^{0}C) in order to prevent the growth of both type E *C. botulinum* and *B. cereus.*

Proteolytic strains of *C. botulinum* (types A and B) do not begin to multiply and produce a toxin until they reach a temperature of 50ºF (10.0^{0}C). Therefore, if produce such as fruits and vegetables, which today are frequently vacuum packaged and contaminated with low levels of *C. botulinum,* are kept at a temperature of less than 50^{0}F (10.0^{0}C), types A and B *C. botulinum* will not cause illness.

Finally, *C. perfringens* is considered to be another control organism. Actually, the highest known growth temperature for a foodborne pathogenic bacterium is that of *C. perfringens* at 126.1^{0}F (52.3^{0}C) and is referred to as the Phoenix phenomenon (164). Therefore, the upper temperature limit for pathogenic microorganism control is 126.1°F (52.3^{0}C) [rounded to 130^{0}F (54.4^{0}C)]. Because of its rapid growth, as rapidly as once every 7.2 minutes at 105.8^{0}F (41.0^{0}C) (196), this pathogen dictates the heating and cooling rates for food. Food must be heated from 40 to 130^{0}F (4.4 to 54.4^{0}C) in less than 6 hours in order to assure no multiplication. Food must be cooled from 130 to 45^{0}F (54.4 to 7.2^{0}C) within 15 hours in order to prevent the outgrowth of *C. perfringens* spores during cooling (98).

Index